VOLCANOES
OF NORTHERN ARIZONA
SLEEPING GIANTS OF THE GRAND CANYON REGION

by Wendell A. Duffield

photographs by Michael Collier

GRAND CANYON ASSOCIATION

Dedication

I dedicate this book to my Mother and Dad, who helped me attain a level of formal education that was not available to them.

Acknowledgments

I owe thanks to two groups of people for making this book happen. First, and highest on my gratitude list, are the audiences for the many volcano talks I've given in Flagstaff and elsewhere. It was, simply put, the remarkable level of interest, attentiveness and curiosity exhibited by these listeners that prompted me to write the book. The settings for the talks have been as diverse as luncheon meetings of service organizations, K-12 classrooms, evening gatherings of tourists around campfires, training sessions for docents of the Museum of Northern Arizona, brown-bagger events of the annual Flagstaff Festival of the Sciences, and a lecture series at the South Rim of Grand Canyon National Park. The general sense of feedback I've felt from audiences is pretty well summed up in a note from a fifth grader in Mrs. Hume's class at Flagstaff's De Miguel School. The student wrote to thank me for my volcano presentation and for "being interesting and not boring, like some of the speakers we had last year." (I offer apologies here to whomever those speakers were....)

For the basic information in the book, I am indebted to a host of fellow scientists, many of whom have studied and written about northern Arizona's volcanoes. I have simply recast the fruits of their research into a style that I hope is understandable and tasty to the nonspecialist.

In alphabetical order, the principals are N.G. Bailey, D.E. Champion, P.E. Damon, W.K. Hamblin, E.W. Hildreth, R.F. Holm, M.A. Lanphere, E.H. McKee, R.B. Moore, C.G. Newhall, E.M. Shoemaker, T.L. Smiley, K.L. Tanaka, G.E. Ulrich, and E.W. Wolfe. To the many other professional colleagues who have contributed to whatever knowledge I have of volcanoes, I also say thanks. You know who you are. And thanks to those who kindly reviewed the manuscript, including Robert Tilling, Kim Watson, Ed Wolfe and Gene Shoemaker. Thanks to my editor, Greer Price, for his dedication to this project from the beginning; to Larry Lindahl for his design; and to Bronze Black for the illustrations.

The color photographs in the book (unless otherwise noted) are by Michael Collier. His aerial shots add a dimension to this book which it would not otherwise have.

My wife, Anne, served as recorder and identifier of geologic features of interest to the non-specialist during trips to create the road logs. She then typed her cryptic and hand-scrawled notes into prose that I could massage. Finally, I thank Anne for her patience as I monopolized the word processor for many a day, and spent time writing that could otherwise have been passed in shared recreation.

Front cover: Sunset Crater, with San Francisco Mountain in the background. Opposite: SP Crater and associated lava flow, under freshly fallen snow. Back cover: SP Crater

Revised 2003
Copyright © 1997 Grand Canyon Association
Text copyright © 1997 Wendell A. Duffield
Photographs copyright © 1997 Michael Collier
Illustrations copyright © 1997 Bronze Black

ISBN 0-938216-58-9   LCN 97-073191

Editor: L. Greer Price
Book design: Larry Lindahl Design
Production supervisor: Kim Buchheit

Printed in China on recycled paper using vegetable-based inks

Grand Canyon Association is a not-for-profit organization. Net proceeds from the sale of this book will be used to support the educational goals of Grand Canyon National Park.

Grand Canyon Association, PO Box 399
Grand Canyon, Arizona 86023
telephone (928) 638-2481
www.grandcanyon.org

Contents

Northern Arizona is internationally famous for the Grand Canyon, one of the world's most spectacular natural features. Far less widely known are the hundreds of geologically young volcanoes of northern Arizona, at least one of which erupted recently enough to bury the homes of local residents. Several of these volcanoes created lava dams within the Grand Canyon not long before humans built their concrete-and-steel counterparts. The volcanic area also includes a volcano whose deformed shape may have been sculpted by an eruption of the sort that violently disrupted Mount St. Helens in the state of Washington in 1980.

Though not as unusual and visually arresting as the Grand Canyon, and therefore not as strong a magnet for attracting visitors, northern Arizona volcanoes are worth knowing about and understanding for a variety of reasons. For example:

◆ They form the hilly, cratered and forested landscape on which much of northern Arizona's population lives, works and plays. Without these volcanoes the area would be a relatively flat plateau, less attractive for human habitation.

◆ The volcanoes have built up the landscape as much as 5,500 feet above a base-level elevation of about 7,000 feet and in so doing have created a variety of local climates and environments. With the higher elevations come many contrasting plant and animal habitats, winter snows, and added recreational possibilities. The highest elevation in Arizona (12,633 feet above sea level at Humphreys Peak) is along the rim of a volcano that is now the site of Snowbowl, perhaps the most exhilarating and challenging of Arizona's down-hill ski areas.

◆ The volcanoes and their eruptive products provide an attractive building stone, cinders for road construction and snowy-day traction, and pumice used in diverse applications, including as a soil improver and as an abrasive to produce distressed-looking, stone-washed blue jeans.

◆ Natural Earth heat trapped in hot and possibly even molten rock at drillable depth within the roots of the volcanoes might someday be tapped for geothermal energy.

◆ There probably will be more volcanic eruptions in northern Arizona.

From the geologic perspective of time, another eruption seems about as likely as another winter snow or summer rain. Of course, the next precipitation is to be expected much sooner than the next eruption. When one compares the average period of volcanic quiet between eruptions (thousands of years) in northern Arizona with the typical human life-span (decades), it's easy to dismiss the idea that anyone living today need worry about the possibility of an eruption during her or his lifetime. Still, as any victim of a so-called "hundred-year-flood" can tell you, nature doesn't always behave in a predictable way. It is simply prudent planning to be aware of the volcanic history and potential of this northern Arizona landscape, so it won't be too much of a surprise when a new volcanic mountain is added to the hundreds already there.

Volcanoes are distributed unevenly all across northern Arizona. I limit my discussion to those of the San Francisco Volcanic Field and others in close proximity to the Grand Canyon, including those that erupted adjacent to the roads that provide access to the South Rim. The volcanoes that line the highways between Williams, Flagstaff, and the Grand Canyon form what geologists call a volcanic field. This term implies a coherent relationship among the field's many volcanoes, both in time and space. This relationship is evident in the digitally produced shaded relief image of the San Francisco Volcanic Field.

Fundamentally, a several-mile-wide swath of volcanism began to develop about 6 million years ago near the present location of the town of Williams. From this starting point, the focus of volcanism has migrated through time east/northeast about 50 miles to its present location northeast of Flagstaff. Geologists refer to this volcanic swath as the San Francisco Volcanic Field, after the largest volcano of the area, San Francisco Mountain. The most recent eruption was only about 930 years ago. The story is by no means complete.

Howard Mesa

Sitgreaves Mountain

WILLIAMS

Mesa Butte Fault

Bill Williams Mountain

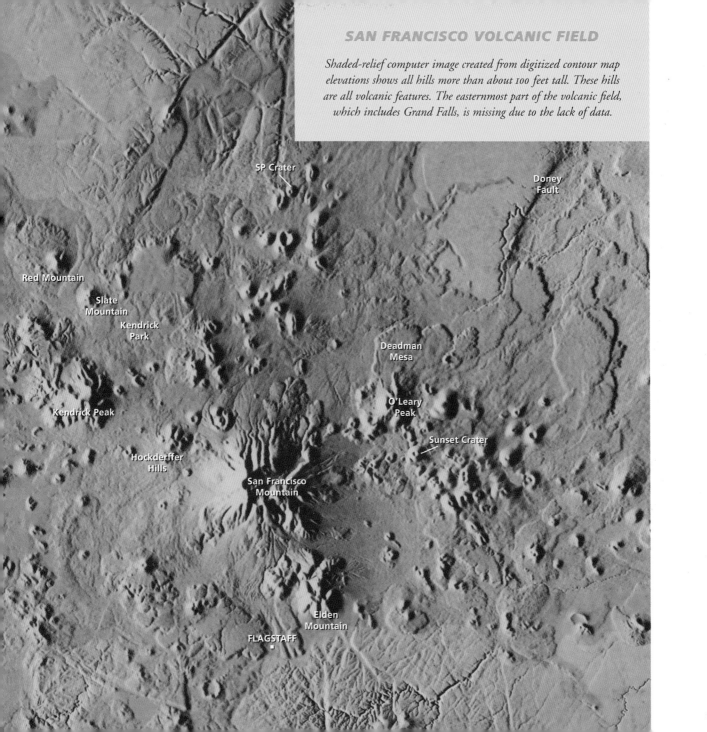

SAN FRANCISCO VOLCANIC FIELD

Shaded-relief computer image created from digitized contour map elevations shows all hills more than about 100 feet tall. These hills are all volcanic features. The easternmost part of the volcanic field, which includes Grand Falls, is missing due to the lack of data.

SP Crater

Doney
Fault

Red Mountain

Slate
Mountain

Kendrick
Park

Deadman
Mesa

Kendrick Peak

O'Leary
Peak

Sunset Crater

Hockderffer
Hills

San Francisco
Mountain

Elden
Mountain

FLAGSTAFF

Chapter 1
The Big Picture

Volcanoes in northern Arizona and elsewhere exist for one very simple reason: they are needed to help the planet Earth cool. Volcanoes throughout geologic time have served this function and will continue to do so until Earth's inner temperature is low enough that the release of thermal energy through eruptions is no longer necessary. Since the Earth has been cooling for about 4.6 billion years, and volcanoes are still a very active part of the cooling process, it seems likely that volcanoes will be around far into the future.

Until it cools sufficiently, the Earth will behave like the hot and restless planet that it is today. Beneath an approximately 50-foot-thick zone at the surface, where ground temperatures tend to change annually with the seasons, temperature increases steadily downward to an estimated maximum of about 5,500° F (3,040° C) in the Earth's core. Part of the core is molten, and

much of the mantle surrounding it is so hot that, within the arena of geologic time, rock there behaves more like a thick and sluggish fluid than a hard and brittle solid.

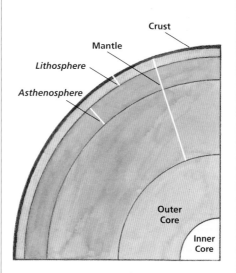

Architecture of the Earth: the traditional, static view widely accepted before the development of plate tectonic theory in the late 1960s.

The Earth's heat originated in part from *planetary accretion*—that is, the gravitational process by which planet Earth formed. Shortly thereafter, thermal energy was added as countless pieces of interplanetary debris collided with the young planet. Most of the physical evidence of such collisions (for example, craters) has been destroyed by weathering, erosion and other surface-modifying processes. By contrast, Earth's moon, unaffected by weathering, erosion and such, clearly preserves a history of intense bombardment on its cratered and pockmarked surface, and thus serves as an example of what Earth also experienced. A third, long-term and ongoing source of Earth's inner heat is the natural radioactive decay of such chemical elements as potassium, uranium, and thorium.

The laws of physics require Earth's inner heat to escape into outer space, as the planet cools. This need for heat to escape is where volcanoes, and other heat-transfer mechanisms, enter the picture. On the one hand, Earth cools

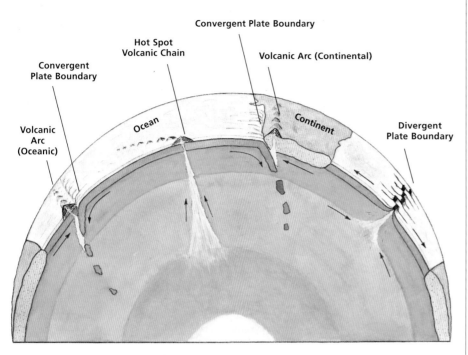

Architecture of the Earth: the dynamic view, although schematic, reflects our current understanding of plate tectonics and convection currents.

as heat moves from the deep interior upward and outward by *conduction* into the atmosphere. Though the daily amount of this heat flow is huge, when integrated over the Earth's entire surface, humans can't feel the effect because it is too low-level and dispersed. Nonetheless, precise measurements of the downward increase of temperature in rocks penetrated by deep drill holes accurately document conductive heat transfer.

Moving heat by conduction is so slow, however, that *convection* and *advection* also are active in order to keep pace with the Earth's need to cool. These "vections" involve the flow of hot material, which moves heat much faster than conduction. Even so, like conduction, convection in the mantle seems inexorably slow, at least within the human-life time frame. Heat transfer by the movement of water and magma in the crust is rapid enough

to be visible daily at many places in the form of flowing hot springs and erupting volcanoes.

PLATE TECTONICS

The crust responds to Earth's cooling by breaking into a dozen or so large pieces that move about much like scum riding on a near-boiling pot of water. The dance of these shifting crustal plates is called *plate tectonics.* The crustal pieces *(lithospheric plates)* move about over a hot and partly molten zone in the mantle called the *asthenosphere* at the rate of about 0.5 to 1 inch per year. These motions don't seem like much unless one thinks in terms of geologic time. For example, during the approximately 6-million-year life-span of volcanism around Flagstaff, the North American plate (which includes northern Arizona) moved westward about 50 miles.

Humans are periodically reminded about plate tectonics by the sudden jerky ground motions of earthquakes, most of which occur along the boundaries between plates. Northern Arizona sits far from such an earthquake-generating zone. The closest plate boundary is the well-known San Andreas fault that slices northwestward through California. Northern Arizona has earthquakes of its own, but these are far less frequent and considerably less powerful than their California cousins.

Like earthquakes, volcanoes also are concentrated near the boundaries between plates. These boundaries are weaknesses in the crust, fundamental physical flaws in the Earth's outer shell that focus the formation and rise of magma to sites of eruption. But every rule has its exceptions, and, as noted above, the volcanoes of northern Arizona are hundreds of miles from a plate boundary. They are thought to lie above an isolated "hot spot" in the mantle, a pipe-like zone where magma is convecting upward from a long-lived source deep within the mantle. Hot spots have been proposed as an explanation for persistent intra-plate volcanism that occurs far from plate boundaries; classic examples of such activity include the chain of Hawaiian islands and the volcanoes that extend from Yellowstone National Park in Wyoming, southwestward along the Snake River Plain.

While tracking the movements of tectonic plates, geologists discovered many examples of lines of volcanoes that are sequentially older along the direction of plate motion. Such a volcanic line, or chain, records a long-lived plume of magma that rises from a stationary source (or hot spot) in the mantle and passes through the overlying, moving plate to a succession of eruption sites.

The most thoroughly documented example of this kind of age progression is the Hawaiian islands, a chain of

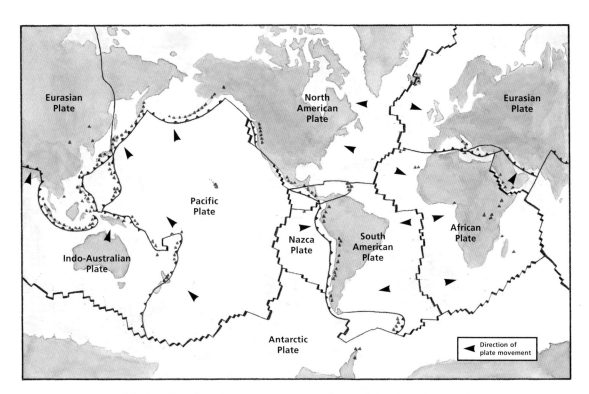

Map of the Earth's surface, showing major tectonic plates and their boundaries. Red triangles indicate zones of volcanic activity. Arrows indicate directions of plate movements.

volcanoes that are incrementally older to the northwest, the direction of motion for the Pacific plate. To the southeast, the youngest Hawaiian volcanoes, Mauna Loa and Kilauea on the island of Hawaii, are still active. The chain continues to grow in this direction, where a submarine volcano, called Loihi, is in the process of building toward sea level, just offshore to the southeast of this youngest island.

Though not as tightly defined as the Hawaiian chain, volcanoes in the vicinity of Flagstaff, Arizona, also exhibit a recognizable age progression. The overall distribution of these volcanoes defines a swath of ground within which bands of restricted ranges of volcano age delineate an eruption path of increasing age to the west-southwest, the direction along which the North American plate is moving. Thus, the center of volcanism has migrated from near Williams, Arizona, to Sunset Crater, northeast of Flagstaff, during the past 6 million years. This 50-mile migration translates into an average motion of about 0.5 inch per year, which is within the range of rates at which all lithospheric plates move. The Arizona chain of volcanoes is simply a somewhat diffuse version of the Hawaiian chain.

A PRIMER ON VOLCANOES

Volcanoes form when molten rock, called *magma*, rises through the Earth's crust and is erupted, gently or

explosively. Magma exists because temperature increases downward within the Earth, and about 20 to 60 miles down, the temperature is high enough to partially melt the rocks found there. This molten material is less dense than the surrounding solid rock and thus, like a cork released under water, rises buoyantly upward into and through the crust to feed volcanic eruptions.

Some magma gets stuck in the crust and solidifies back to rock while still underground. Such rock is called *intrusive* (or *plutonic*) and is available for study only after erosion exposes it at the surface or when sampled by drilling. Volcanic rocks are called *extrusive,* because the magma in this case extrudes onto the Earth's surface. A vent marks where a magma-carrying conduit intersects the Earth's surface. A volcano is the landform that builds up around a vent, the place where magma extrudes. *Lava* is the term geologists use to refer to magma which has erupted onto the Earth's surface. A complete definition of a volcano should also include reference to the magma-carrying conduit as well as the overlying landform.

Volcanoes come in a variety of sizes and shapes, and their behavior ranges from quietly benign to violently destructive. Why such diversity? The

Unnamed cinder cone 8 miles east of Sunset Crater. Its shape is clearly indicative of the youthfulness of the eastern part of the San Francisco Volcanic Field.

answer to this question is fundamentally related to the chemistry of the erupting magma. Chemical compositions that produce thin, easy-flowing magmas are associated with nonviolent eruptions, whereas compositions that produce thick, sluggish-flowing magmas correlate with explosive volcanic events.

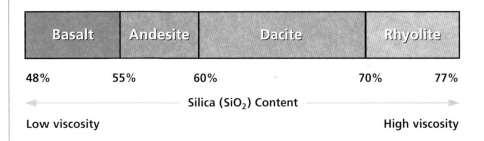

Common magma types, their composition, and viscosity.

With rare exception, magma is a *silicate liquid:* the most abundant chemical building block is a pyramid-shaped assembly of the element silicon (Si) surrounded by four oxygen (O) atoms. Other elements that are common rock-forming materials (including aluminum, iron, magnesium, potassium, sodium, and calcium) occupy the spaces between and around the silicate building blocks. The overall mixture forms the hot and sticky stuff called magma.

With different relative amounts of these chemical constituents, magma takes on a spectrum of chemical compositions. Magma that forms in the mantle carries the name basalt. Though fairly uniform in composition,

even this "mother" magma is somewhat variable; much of the variation in composition depends upon what happens to basalt magma once it leaves the mantle and starts its upward journey through the crust. For example, rocks encountered in the crust sometimes are partly melted by, and dissolved into, rising mantle-derived magma, and sometimes crystals grow and then separate from magma as it begins to cool. Both of these processes usually result in an increased silicon content for the modified or "daughter" magma, and thus the natural tendency is for basaltic magma to change toward a more silicic composition. How far such change advances depends on the duration of the magma journey and the chemical character of the crust traversed.

Magma composition is classified and named principally on the basis of the amount of silicon, which rock analysts traditionally report as silicon dioxide (SiO_2) and call *silica.* The chart above summarizes names of the common magmas and their associated ranges in silica. Magma names double as the names of volcanic rocks that form from those magmas.

Most of the volcanic rocks of northern Arizona are within the silica-poor end of the compositional spectrum and thus are called basalt or andesite. Relatively silica-rich rocks, dacite and rhyolite, are present but are common only locally around Flagstaff.

One of the most important properties of magma, and one that determines eruption style and the eventual shape of the volcano it builds, is its resistance to flow, called *viscosity*. From everyday experience we know that liquids with low viscosity, such as milk, flow immediately from the jostled glass, whereas a more viscous liquid, such as maple syrup, may not leave its overturned container until the pancake is cold.

Even the most fluid of magma, basalt, is far more viscous than such syrup, and magma viscosity increases as its silica content increases. This causes rhyolite, the most silica-rich of all magmas, to pile up in sticky masses right over its eruptive vent to form tall, steep-sided volcanoes. At the opposite end of the viscosity spectrum, basaltic magma typically flows great distances from its eruptive vent to form low, broad volcanic features. Toward the center of the viscosity spectrum, andesitic magma tends to produce volcanoes with profile shapes between these two extremes.

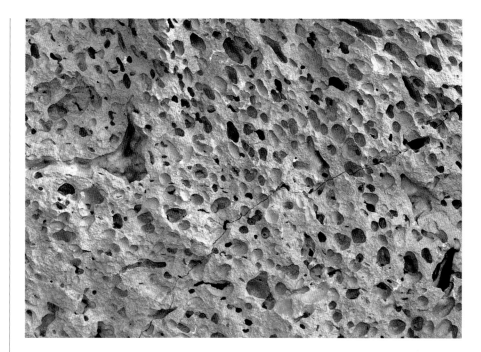

Vesicular basalt. These cavities—generally smaller than $1/4$ inch—result as gas, trying to escape from molten lava, is trapped during solidification.

Water is an additional important magma ingredient. Carbon dioxide (CO_2) and various sulfur-containing gaseous constituents are generally also present in magma (in solution) but usually at concentrations that are ten or more times lower than that typical of water. These substances are grouped under the term *volatiles* because they tend to occur as gases at temperatures and pressures common at the Earth's surface.

As basaltic magma changes composition toward rhyolite, volatiles become concentrated in the silica-rich magma.

When contained in sufficiently high concentration, volatiles (principally water) produce highly explosive volcanic eruptions. Whatever their concentration, however, volatiles can only be held in magma by confining pressure. Otherwise they come out of solution and escape as gases. A similar thing occurs when you remove the top from a bottle of soda pop and reduce the pressure on the liquid inside: the carbon dioxide in solution then escapes as gas bubbles.

Within the Earth, confining pressure is provided by the load of overlying rocks. Such a rock cover serves the same function as the firmly attached lid on a home pressure cooker. As magma rises from the mantle to depths of about a mile or somewhat less, however, the rock load is reduced to the point where volatiles, principally water, start to boil off. Even closer to the surface and during eruption, the rock load is close to zero and water can boil off more easily—not too easily, though, because viscous drag on rising bubbles inhibits escape of the water vapor. The higher the viscosity of the magma, the greater the inhibiting force.

Try blowing bubbles through a straw dipped into a glass of milk and then into a jar of maple syrup. Bubbles rise rapidly through the milk and can even produce a fine milky spray in your face as they escape quickly into the atmosphere. Bubbles rise more slowly through the syrup. Some may even remain trapped if the syrup is cold enough.

Bubbles of water vapor rising through basaltic magma escape easily enough during eruptions to propel a fine aerosol-like spray of magma called a lava fountain. Eruptions of Hawaiian volcanoes typically begin with this type of fountain, because all Hawaiian

magmas are basalt. Though sometimes as tall as 2,000 feet and impressively powerful in their own right, fountains of basaltic magma pale to wimpishness next to their rhyolitic counterparts. Bubbles rising through highly viscous rhyolitic magma have such difficulty escaping that many carry blobs of magma and fine bits of rock with them when they finally break free and jet violently upward to feed a hot and therefore buoyant eruption column that can rise miles above the Earth. Once dispersed within the upper atmosphere, fine volcanic debris of such a powerful eruption column screens out sunlight that otherwise would reach the Earth's surface, and thus causes widespread, if not global, changes in the weather. The 1982 eruption of El Chichón Volcano in Mexico and the 1991 eruption of Mount Pinatubo in the Philippines each produced such an effect. The greater the original gas concentration in a magma and the greater the volume rate of magma leaving the vent, the taller the eruption column that is produced.

Gases that escape from magma during eruption are mixed into the atmosphere and become part of the air

that humans, other animals, and plants breathe and assimilate. However, as magma cools and solidifies to rock during eruption, some of the gas that is trying to escape remains trapped as bubbles, called *vesicles*. Almost all volcanic rocks contain some gas bubbles. Pumice, a variety of vesicular rhyolite, is mute testimony to magmatic gas that never got away. Some pumice is so vesicular that the rock floats in water. Cinders are simply the basaltic equivalent of pumice.

With regard to the possibility of future gas-driven explosive volcanic eruptions in northern Arizona, the good news is that almost all past eruptions have been relatively non-explosive. Most, in fact, have been of the quiet basaltic-magma variety. So, if the past volcanic history is a dependable guide to what the future holds, chances are that the next eruption will be similarly non-explosive.

Volcanic Features of Northern Arizona

The San Francisco Volcanic Field covers an 1,800-square-mile area centered roughly on Flagstaff, Arizona. Of the literally hundreds of volcanoes that decorate the landscape throughout the region, almost all are what geologists call *cinder cones* (also known as *scoria cones*).

CINDER CONES

At least 600 cinder cones have been mapped, counted and catalogued in northern Arizona. Geologists have been so overwhelmed by the sheer abundance that, to avoid confusion, they unambiguously identify each volcano by number rather than by name. Many of the prominent features bear names, too, and so the cinder cone assigned #3824 is more widely known as Sunset Crater, #3036 as Merriam Crater, #5703 as SP Crater, and so forth.

Cinder cones grow when basaltic magma erupts in fiery orange and red fountains of lava spray that fall back to Earth, building a cone-shaped or round-topped hill around the erupting vent. Each individual cinder is simply a lava droplet or blob that solidifies before

falling back to Earth. Blobs that are still molten when they fall back may feed lava flows. A lava flow may also form when magma of low volatile content quietly oozes from a vent, without fountaining. SP Crater, located about 26 miles north of Flagstaff, is a nearly

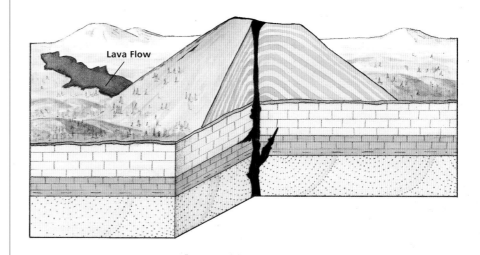

Lava Flow

Internal structure of a typical cinder cone.

perfect example of a cinder cone and its associated lava flow. In this instance, the flow spread outward and downslope over a landscape dipping gently to the north (see photo on page 58).

Another noteworthy cinder cone and associated flow were produced by eruption from a vent marked by the volcano called Merriam Crater, about 20 miles east of Flagstaff. This flow moved down a gentle eastward-sloping surface for about 7 miles, where it cascaded into the 200-feet-deep, steep-walled valley of the Little Colorado River and subsequently advanced an additional 15 miles downstream. Today, the entry point into the Little Colorado River is marked by the cliff-forming Grand Falls (Road Log, Leg 7, page 59).

Hundreds of other cinder cone/lava flow pairs make up the hilly landscape roughly bounded by a line drawn from Flagstaff to Grand Falls to Cameron to Valle to Williams and back to Flagstaff. Simply count the hills in this area and you'll know there are at least that many volcanoes.

Each cinder cone has a characteristic internal structure expressed as layer upon layer of cinders dipping outward radially from the eruptive vent. An individual layer may be as thin as a few inches or as thick as several feet. Layering results from fluctuations of eruptive intensity; a steady eruption would produce a massive nonlayered accumulation of cinders.

Most cinder cones in northern Arizona are too young for erosion to

have exposed their internal structure. Cinder layers are exposed in the walls of some quarries around Flagstaff, but sets of parallel marks from the scraping of mining machinery can sometimes be misidentified as layering by the trained and untrained eye alike. The best natural exposures of cinder layers within a cone near Flagstaff are at Red Mountain, about 26 miles northwest of

One of the best known features of its kind in the world, Meteor Crater is not a volcanic feature at all, but rather an impact crater, the result of a meteorite collision that occurred about 50,000 years ago. In the distance are San Francisco Mountain and other smaller neighboring hills, all volcanic. U.S. Geological Survey photo by Dave Roddy.

Flagstaff along Highway 180 (see page 50). Erosion has carved a bowl-shaped hollow out of the northwest flank of this volcano, exposing a sequence of gently north-dipping layers, which can be seen from the highway and (for those who would like a closer look) along a marked pathway that leads into the eroded-out bowl itself. The Red Mountain cinder cone is extremely asymmetric in shape: a lopsided volcano that apparently is the product of eruption during a time of wind out of the southwest. A strong, steady wind during lava fountaining results in the preferential deposition of cinders downwind of the vent, whereas calm conditions favor formation of a symmetrical cone centered around the vent. Rapidly changing winds can produce both kinds of cinder deposits during a single eruption. The northeast side of the Red Mountain cone is about 900 feet tall. So few cinders accumulated on what is inferred to have been the upwind side of the erupting vent that the southwest part of the cone is barely evident in today's landscape. By contrast, SP Crater, Merriam Crater, Sunset Crater, and most of the other cinder cones of the region, developed almost symmetrically around their vents.

The sheet-like lateral extent of a lava flow associated with a cinder cone principally reflects the high fluidity (low viscosity) of basaltic magma. Anyone who has witnessed an active basalt flow, like those at Kilauea Volcano in Hawaii, has seen this type of lava

move over the landscape much like water. Additional factors influence the thinness and lateral extent of a basalt flow. Other factors being equal, the configuration of the land surface over which magma is erupted can produce flow shapes ranging from a thick puddle (in a local closed basin), to a broad thin sheet (over a gently inclined planar surface), to a very long thin finger (along a well-defined river valley). Moreover, magma that gushes rapidly and copiously from a vent will produce a much more extensive flow than magma oozing out slowly.

Whatever the lay of the landscape and the volumetric rate at which magma is erupted, flowout distance is limited by the interplay among these factors and the rate of magma cooling. After all, a molten substance that solidifies to rock at about 2,190° F (1,200° C)

The lava cascades and cinder cone at Toroweap, in the western Grand Canyon, are analogous to the 1969 eruption of Kilauea Volcano, on the island of Hawaii. Right: U.S. Geological Survey photo.

cannot long remain fluid at the Earth's surface. Lava flows emplaced over relatively flat ground in the Flagstaff area are typically a mile or so wide and tens of feet thick.

For obvious reasons, lava that follows a river valley is described as an ***intracanyon flow***. In addition to the intracanyon flow that formed Grand Falls along the Little Colorado River, about a dozen fascinating intracanyon flows partly filled the Grand Canyon during a 1.4 million year period in recent geologic times. Geologists have also mapped a basalt flow that followed

a stream valley for almost 45 miles, from near Kendrick Peak to Cameron—a remarkable distance that suggests an unusually high rate of magma effusion at the vent.

LAVA DAMS IN THE GRAND CANYON

After considerable study, geologists have concluded that lava flows dammed the Colorado River in the Grand Canyon at least thirteen times since about 700,000 years ago. The most recent dam formed about 400,000 years ago or perhaps even more recently—a twinkling of an eye for a planet 4.6 billion years old. All dams were built along the stretch of the river between Toroweap Valley and Whitmore Canyon, 178 and 188 miles downstream, respectively, from the zero reference marker at Lees Ferry.

A Closer Look:
Grand Falls

When humans build a dam, the reservoir behind it has a finite, useful lifetime. Eventually, the unending processes of erosion and deposition fill a reservoir. With sediment fill, useful functions of a dam (power generation, flood control, and recreation) are lost, even though the dam itself remains sound. The reservoir behind Glen Canyon Dam at the Arizona-Utah border is thought to have a useful lifetime of 300 to 500 years. It is difficult to be more precise, because the rate at which sediment is washed in is not accurately known.

An upstream dam can extend the useful life of one downstream by trapping sediment that otherwise would move down the drainage system. For example, when Glen Canyon Dam was built at the upstream end of Grand Canyon, it extended the life of Hoover Dam at the downstream end.

Reservoirs created by lava flows that pond in river canyons also have finite lifespans, which are different for different types of rivers. For example, when a lava dam and associated reservoir form in the canyon of a *perennial* river, the lifespan of the pair is limited by erosion of the dam—a process that proceeds rapidly once water begins to spill over the top of the dam. If, however, water flow in the river is *intermittent*, spillover may rarely or never occur and the lifespan of the reservoir, like those behind humanmade dams, is limited by the time needed to fill with sediment.

Spillover and water erosion were limiting factors for lava dams and associated

Grand Falls

reservoirs in the Grand Canyon (see *Lava Dams in the Grand Canyon,* page 15) because the Colorado River flows year round. In contrast, sediment fill ended the life of the reservoir created when a lava flow dammed the valley of the Little Colorado River at the place now called Grand Falls, because the Little Colorado River flows intermittently.

The dam-building lava flow at Grand Falls erupted from either the cinder cone known as Merriam Crater or a smaller volcano nearby, about 7 miles west of the Little Colorado River. Geologists do not agree on exactly when this eruption occurred, but it may have been several thousands or tens of thousands of years ago. As the flow spilled into the river, it filled the entire 200-feet-

deep valley at the point of entry and spread a short distance over the east bank. A downstream-tapering finger of lava flowed along the river valley for about 15 miles beyond the point of entry.

Sediment began to accumulate behind the new lava dam with the next storm that produced runoff upstream. Eventually, the reservoir behind the dam filled with sediment, and all subsequent river flow followed the low ground around the east edge of the lava, along its contact with the relatively soft sedimentary rocks of the plateau east of the river. Thus, Grand Falls initially developed, and has continued to evolve in sedimentary rocks, not lava.

The volcanoes that produced the dams are seen today as basaltic cinder cones and associated lava flows. These cones tend to occur along north-south lines, because they formed where magma rose along deeply penetrating fissures of this orientation. Most of the eruptions originated somewhat north of the canyon and fed flows that cascaded southward into the Grand Canyon gorge as incandescent lava falls. Remnants of frozen lavas that locally blanket canyon walls are mute reminders of these fiery cascades (see photo on page 15).

Erosion has removed most of the evidence of the lava dams themselves. What remains is clearly visible as bits and pieces of horizontal lava flows banked against the walls of the canyon. Where visible, these flows display classic columnar jointing. Upstream, deposits of silty and sandy sediment that typically settle out on any lake bottom are locally preserved on the canyon walls at elevations high above present river level, and in tributary valleys. One of the few places where such sediments are preserved is Havasu Canyon.

Taken together, the lava dam and sediment remnants record a series of past lakes within the Grand Canyon. Reconstruction of a highly accurate picture of the succession of dams and their lakes from just the few available pieces of the puzzle is challenging. But even if some fine texture remains blurred, the general scene is quite recognizable. The following sequence

of events was repeated from episode to episode of dam formation.

Eruption produced a lava flow or flows that partly filled the Grand Canyon. Some flows advanced as far as 85 miles downstream from their entry points. Upstream, water ponded against the new lava dam, forming a

Columnar-jointed basalt, in the western Grand Canyon. These fractures or joints are due to shrinkage that occurs as lava cools and solidifies.

lake. Rising lake level, fed constantly by the upstream river system, eventually overtopped the lava dam, and spillover began erosive downcutting. Without the presence of a carefully engineered and solidly constructed spillway, the

dam was quickly removed, reestablishing a pre-dam profile for the river and resetting the scene for another dam/lake/erosion episode. The principal variation on this overall theme was in the height of the dam, which controlled the maximum depth and upstream length of the associated lake.

The highest-elevation remnant of the biggest lava dam is about 2,300 feet above present river level. From bottom to top, this lava dam was several hundred feet taller than Glen Canyon Dam at Page, Arizona. Lake water impounded behind this lava dam extended upstream through the Grand Canyon into Utah nearly as far as present-day Moab. Below the canyon rim at what is now Grand Canyon Village (river mile 90), the lake was about 1,600 feet deep. Calm-water recreation possibilities would have been plentiful, though river running was out of the question.

The highest-elevation remnant of the smallest lava dam is only about 200 feet above present river level. The lake behind this dam extended only about 37 miles upstream, considerably downstream from Grand Canyon Village. In the intermediate size range, several of the dams backed up lake water as far as Grand Canyon Village.

If one knows the volume of a reservoir and the average rate at which water flows in from the upstream catchment basin, the time required to fill the reservoir to capacity can be calculated. Though the inflow rate must have varied during the next period when lava dams were formed, filling times probably ranged from several years to a few days, for the entire spectrum of dam sizes.

The time needed for erosion to effectively remove a lava dam is more difficult to define, but it may have ranged between 100s and 1,000s of years, based on the rate at which a falls like Niagara retreats upstream by erosion. Keep in mind, though, that a lava dam is essentially a long, downstream-sloping wedge.

If a lava dam were shaped like its thin humanmade counterparts, Hoover Dam and Glen Canyon Dam, removal by erosion probably could be accomplished much more quickly, in a few years or decades. Whatever the time frame, failure of the relatively large lava dams probably produced brief, spectacularly powerful, and damaging downstream flooding.

Aside from the differences in profile shape and construction materials, a lava dam contrasts fundamentally with the human-made ones because it has no means of releasing water downstream in a way that will not harm the dam. Once spillover begins, destruction of a lava dam almost certainly proceeds rapidly along side contacts with soft sedimentary rocks and along the base of the dam where a churning plunge pool abrades and carries away pieces of the basalt rock and its underpinnings.

During the spring of 1983, upstream runoff from the Colorado and Utah parts of the Colorado River

basin was so voluminous that the rate of water released from Lake Powell could not keep pace with the rate of upstream inflow. As a result, the spillway for Glen Canyon Dam became active for only the second time since the dam was built. Despite the great care and thought that had gone into design and construction of the spillway, considerable damage occurred as water surged through the spillway conduit. Steel-reinforced concrete lining was torn away and house-size cavities were eroded into the relatively soft adjacent sedimentary rocks. Fortunately, the need to continue the water release through the spillway vanished before the situation became critical. Still, it is sobering to think that such a massive, human-made, concrete-and-steel plug might have failed rapidly by erosion, much like a lava dam.

Imagine the dilemma over what to do if volcanic eruptions produced another lava dam in the Grand Canyon, now that many people and much property are at risk from the ensuing catastrophic flooding. When volcanic debris associated with the 1980 eruption of Mount St. Helens (southwestern Washington) formed a natural dam in the Toutle River drainage, the U.S. Army Corps of Engineers was called upon quickly to build a spillway to control outflow and prevent catastrophic failure of the new volcanic dam. In theory, this solution could also be applied should a new volcanic dam form in the Grand Canyon. Unlike the

Mount St. Helens experience, however, access to the construction site would be extremely difficult.

SHIELD VOLCANOES

When a vent produces many successive basaltic lava flows stacked one on top of another in eruptive order, the resulting landform is called a *shield volcano*. A cinder cone and its associated lava flow can be thought of as the initial building blocks of a shield volcano.

A cinder cone is *monogenetic*, because it forms from a single, short-lived eruption (of a few to a couple of hundred years duration). In contrast, a shield volcano, and any other volcano that is an accumulation of the products of many eruptions over a period of, say, thousands to hundreds of thousands of years, is *polygenetic*.

The world's largest shield volcano, Mauna Loa in Hawaii, is 150 miles wide and rises about 5 miles up from the floor of the Pacific Ocean. All of the Hawaiian islands are built of shield volcanoes. By comparison, northern Arizona is home to only a handful of volcanoes that might be classified as shields, and most of these are true miniatures, less than 3 miles wide and no more than 1,000 feet tall. Perhaps the best-developed local example, the Hart Prairie shield, is centered at Hart Prairie, along the west-sloping lower flank of San Francisco Mountain. Some lava flows of this rather flat-topped volcano extend at least 10 miles to the west and southwest. The original extent is unknown, because flows from younger volcanoes have buried much of the Hart Prairie shield. The shield could not grow much to the east, because the underlying west-sloping flank of San Francisco Mountain was an insurmountable barrier.

Why did so few of the hundreds of basaltic volcanoes in northern Arizona grow beyond the monogenetic cinder cone stage? The answer is unclear. For some reason, many small-volume batches of magma were erupted at hundreds of vents, rather than a few large volumes being concentrated at just a few vents. One can speculate that magma rose from a broad, rather than a tightly focused, hot-spot source zone in the Earth's mantle. In addition, perhaps a complex and pervasive network of crisscrossing fractures and other flaws in the 30-mile-thick sequence of crustal rocks provided many ready-made pathways of diverse orientations to disperse magma laterally as it worked its way upward toward sites of eruption.

LAVA TUBES

Sequences of stacked-up basaltic lava flows almost invariably contain a subterranean system of open passageways called lava tubes. A well-developed system includes tubes at various depths leading to a variety of destinations, volcanic metro lines that may be thought of as Vulcan's subway. Because most eruptions in northern Arizona

produced basaltic lava flows, Vulcan's subway system is well developed near Flagstaff.

A lava tube begins to form when a channel carrying lava becomes crusted over with solid rock, as the still-molten lava continues to flow beneath that crust. The chilling effect of the atmosphere on lava flows can produce such a crust within minutes. (Basaltic lava solidifies at about 2,190° F/1,200° C and a hot summer day in Flagstaff is only

about 95° F/35° C.) Once formed, a tube can carry flowing lava for months to years, so long as lava is continuously supplied to the upstream end of the tube. The insulating effect of the solidified rock keeps lava inside the tube at a temperature above its solidification point for long periods of time.

The process of a lava stream roofing over to form a tube is akin to what happens when a river ices over in winter. But while the roof of a lava tube remains solid rock "forever," the icecap on a river melts as spring weather arrives. Moreover, rivers capped with ice don't normally result in open tubes, because a copious and continuous upstream supply of water keeps the tube full. However, as a volcanic eruption winds down and eventually stops, a roofed-over stream of lava partly or completely

Active and ancient lava tubes, in Hawaii (top) and the San Francisco Volcanic Field, respectively. U.S. Geological Survey photos.

drains out the downstream end of the tube, leaving an open subterranean passageway.

Lava tubes that do not completely drain have relatively flat floors made of "frozen" residual molten lava. Pre-existing tubes may be reoccupied by lava from later eruptions that may build up the floor. Wall decorations above a tube floor sometimes include bath-tub rings of solidified basaltic ledges that formed as the lava level dropped incrementally.

Jumbled piles of rocks on tube floors usually indicate local collapse of the roof. If such collapse works its way to the Earth's surface, a *skylight* is formed. Pieces of collapsed roof that fall into flowing lava may be preserved as tilted and generally out-of-place looking slabs set in an otherwise uniformly textured flat lava surface. Like their limestone counterparts, lava tube caves sometimes have stalactites dangling from the roof and, very rarely, stalagmites sticking up from the floor. These decorative features are usually only inches long at most and are made of basaltic lava instead of the mineral calcite, the building block of limestone.

The roofs of many lava tubes are coated with a smooth glaze that represents a kind of volcanic-rock ceramic resulting from prolonged exposure to near-melting temperatures. A tube that actively carries lava long enough (on the order of weeks to months) can heat its floor to the melting point and thereby "erode" downward.

Some lava tubes are smaller than the human body, but many are large enough for human exploration. Spelunkers and others who map lava tubes sometimes find complexly shaped and branching passageways, much like the form of a braided river bed. Lava tubes may exist at multiple levels, the result of many separate tube-forming lava flows being stacked up to form a lava field. Even a single continuous tube may abruptly change levels, presumably because it mimics vertical relief on the ground surface over which it originally formed.

Like other natural underground passageways, lava tubes are discovered as a result of a diligent well-planned search, or by accident. The easiest tube to discover is one whose roof has collapsed, providing easy access to the explorer through a skylight. In Hawaii, where the islands are built almost entirely of stacks of basaltic lava flows riddled with lava tubes, some are accidentally discovered when heavy equipment breaks through the roof of an otherwise hidden passageway. Lava tubes as long as 34 miles have been charted in Hawaii.

Perhaps the most widely known lava tubes near Flagstaff are those at Sunset Crater Volcano National Monument, about 15 miles northeast of town, and at Lava River Cave (sometimes known as Government Cave) about 20 miles north-northwest of town in the Coconino National Forest. The tubes at Sunset Crater are closed to the public, primarily for safety reasons. Nearly a mile of Lava River Cave is open to the public, and several of the features described above can be seen there. This tube formed in a lava flow erupted from the Hart Prairie shield volcano described earlier. (For additional information about how to get there and what you can expect to see in this tube, contact the Peaks View Ranger Station, Coconino National Forest, in Flagstaff at 928-526-0866.) If you decide to explore, be sure to take a warm coat and a reliable flashlight, your tickets to safe travel on Vulcan's subway.

STRATOVOLCANOES

A *stratovolcano* (also known as a *composite volcano* or *stratocone*) is an upward-steepening, sharp-peaked mountain whose profile most people identify as the classical volcanic shape. Mount Fuji, in Japan, is the best-known example of a stratovolcano. In the United States, the pre-1980 shape of Mount St. Helens in Washington state is perhaps the most widely known example, simply because of the violent 1980 eruption that spread ash across several states to the east and provided an in-your-face reminder that active volcanoes do exist in the United States outside of Hawaii and Alaska.

San Francisco Mountain, the single stratovolcano in the San Francisco Volcanic Field.

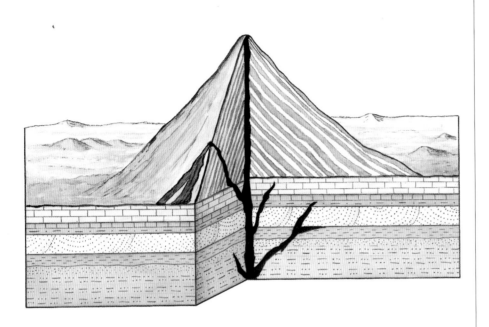

Internal structure of a typical stratovolcano.

The Flagstaff area's one and only strato-volcano is San Francisco Mountain, the lofty landmark at the north edge of town which includes Humphreys Peak, Arizona's highest peak at 12,633 feet.

Whereas a cinder cone is built of layers of basaltic cinders and a shield volcano is a stack of lava flows, a strato-volcano consists mostly of alternating layers of loose volcanic fragments and coherent lava flows. The layers of frag-ments accumulate as fallout from lava fountains and taller eruption columns (akin to the formation of cinder cones), and as fast-moving avalanches of hot fragments and gas sporadically derived from these fallout deposits. Fragmental layers are also formed when loose vol-canic debris are mixed with water from snowmelt or rain on steep volcano flanks to produce slurries (mudflows or *lahars*) that move swiftly down moun-tainsides and valleys.

Stratovolcanoes are the accumu-lated products of many eruptions and lahar-forming events that occur over tens of thousands to as many as a half-million years. Chemically, most of these products are andesite; some are dacite; and a few are basalt and rhyolite. This chemical mix and the characteristic interlayering of lava flows and fragmen-tal deposits are aptly described by the stratovolcano synonym, composite volcano. The classic Mount Fuji profile of many stratovolcanoes is interrupted by steep-sided "parasitic" volcanic hills that form when a small amount of thick and sticky dacitic or rhyolitic magma oozes out of fissures near the top or lower on the flanks of the larger host volcanic edifice.

The characteristic internal structure of a stratovolcano is layered, a shingled arrangement of deposits that dip out-ward radially from a central eruptive vent. This volcanic architecture is readily apparent at San Francisco Mountain, because the innards are well exposed in what is called the Inner Basin. This basin formed when northern Arizona's lone stratovolcano lost its pristine Mount Fuji shape, hundreds of thousands of years ago, during a col-lapse of the flank perhaps like the one that occurred during the 1980 eruption that sculpted Mount St. Helens into a similar shape.

The truncated edges of deposits that once extended upward to the vol-cano's summit form a series of parallel ribs around the walls of the Inner Basin. Reconstruction of the original extent of these sheet-like building blocks indicates that the pre-basin summit was about 16,000 feet above sea level—an impressive landmark indeed. About 400,000 years ago, as this stratovolcano grew to its maximum height, the rocks that form the six prominent peaks (Reese, Aubineau, Humphreys, Agassiz, Fremont and Doyle) around the rim of today's Inner Basin were 3,500 to 4,500 feet down the flanks of the mountain.

SAN FRANCISCO MOUNTAIN: A PREHISTORIC VERSION OF MOUNT ST. HELENS?

San Francisco Mountain once towered nearly 9,000 feet above its surroundings. It is northern Arizona's version of Mount Fuji, except that the original peak and part of the core of the volcano are missing, as though a giant ice cream scoop cut away the top and east side of the mountain. The how and why of these missing parts have long been a mystery to geologists. Recently, the eruption of another stratovolcano, hundreds of miles away, provided at least one viable explanation.

Prior to 1980, Mount St. Helens, in Washington state, was an almost perfectly shaped stratovolcano that rose 6,000 feet above its surroundings. Suddenly, this beautifully symmetrical shape began to deform into a menacing monster. In late March 1980, many earthquakes began to originate beneath the volcano. These shook the mountain and surrounding area repeatedly. Within a week of the onset of earth-

The eruption of Mount St. Helens, a stratovolcano, in Washington state in 1980. Photos by Keith Ronnholm.

shaking, a series of steam explosions blasted two large holes through ice and rock at the summit, and simultaneously a blister-like bulge began to grow on the north flank as magma intruded into the volcano.

This behavior of shake, blast and bulge continued until 8:32 AM (Pacific Daylight Time) on May 18, when most of the north flank, precariously over-steepened by the bulging, collapsed. This collapse triggered a powerful north-directed release of pent-up gas contained in magma and hot water that had been sealed within the volcano. When the air finally cleared, observers could see that the uppermost 1,300 feet of the stratovolcano had disappeared and a north-facing amphitheater-like crater had formed. Nearly 250 square miles of mature Douglas-fir forest were flattened by the force of the eruptive blast directed northward from the heart of the volcano. Then, during the following months to a few years, almost as an afterthought, a lava dome grew in the crater. The volcano today is apparently dormant, although additional eruptive rumblings would not be unexpected.

Geologists quickly applied lessons learned at Mount St. Helens to stratovolcanoes elsewhere, especially those with large open amphitheaters on the mountainside. The origin of these was puzzling. The force and power of a violent volcanic eruption is commonly directed upward (vertically rather than laterally), giving rise to an evenly

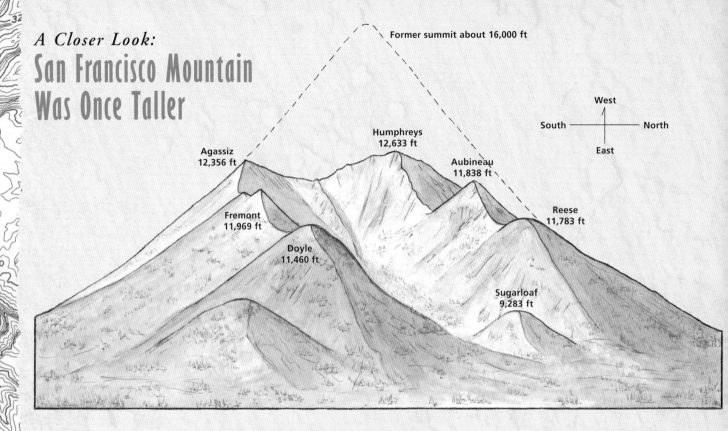

A Closer Look:
San Francisco Mountain Was Once Taller

Former summit about 16,000 ft

West
South — North
East

Humphreys
12,633 ft

Agassiz
12,356 ft

Aubineau
11,838 ft

Fremont
11,969 ft

Reese
11,783 ft

Doyle
11,460 ft

Sugarloaf
9,283 ft

The original shape of a truncated strato-volcano, such as San Francisco Mountain, can be reconstructed from the configuration of what is left of the lava flows and cinder layers that are building blocks of the volcano. This kind of exercise is the fundamental stuff of the geologic profession—creating a complete picture when only some pieces of the puzzle are available.

The truncated edges of outward-sloping lava flows and cinder layers are exposed around the rim and in the walls of the Inner Basin. The direction and inclination of these deposits can be measured and then projected upward. When this is done, the projected layers intersect at a point that can

be inferred to represent the summit of the undeformed volcano. If anything, this underestimates the elevation of the original summit, because volcanic deposits of a stratovolcano tend to be more steeply inclined towards the top.

The reconstructed shape of San Francisco Mountain indicates a former summit at about 16,000 feet above sea level. The high point presumably existed about 400,000 years ago, the youngest known age of the lava flows that make up the body of the volcano. Today's highest elevation on the mountain, Humphreys Peak at 12,633 feet, can be seen from vantage points one hundred or more miles

away. The earlier, higher version of the volcano would have been visible from even greater distances.

The highest elevation in the forty-eight contiguous states today is Mount Whitney in California at 14,495 feet. It's tempting to think that Arizona's San Francisco Mountain was once the highest point in the lower forty-eight states. However, Mount Whitney and other high peaks of the mountainous American West also may well have been taller than the San Francisco Mountain stratovolcano 400,000 years ago; all of these mountains have experienced an unspecified (but considerable) amount of erosion since then.

developed closed crater around the vent. Mount St. Helens, however, showed that volcanic fury can be directed horizontally, hollowing out an open-sided crater. Thus, the long-standing mystery of how San Francisco Mountain lost its top, core and east flank suddenly had a new explanation for geologists to consider.

The general shapes of the two volcanoes are quite similar, although the basin at San Francisco Mountain, called the Inner Basin, is notably larger. Like post-1980 Mount St. Helens, San Francisco Mountain stratovolcano also has a lava dome, called Sugarloaf, that grew after the amphitheater basin formed. During the pre-1980 years, geologists who studied San Francisco Mountain were unable to develop a viable explanation for the decapitation and gutting of the volcano. Several speculative hypotheses were suggested, none totally satisfying: perhaps magma within the volcano was suddenly with-drawn downwards or sideways, causing rocks to collapse into the space the magma once occupied; maybe "normal" water erosion removed that flank; perhaps the basin was carved out by glaciers.

In light of the events at Mount St. Helens, the list of maybes grew to include the possibility that magma had intruded into the volcano and, in so doing, pushed the east flank outward until it was so steep it collapsed from the pull of gravity, perhaps initiated by an earthquake. Such collapse would have triggered an east-directed gas explosion as pressure was released on

The shape of the recently erupted Mount St. Helens (above) is remarkably similar to the shape of San Francisco Mountain (below). U.S. Geological Survey photos.

magma and hot water contained within the stratovolcano. The list of maybes is long, and the answer is still not clear.

What we do know is that the Inner Basin formed after the outermost lava flows had been added to the flanks of San Francisco Mountain and before the Sugarloaf lava dome grew. Radiometric dates from these rocks indicate that the basin formed between about 400,000 and 200,000 years ago. Given northern Arizona's climate, both at present and back to the basin-forming time, it seems reasonable that mountain streams and glaciers have modified the shape of the Inner Basin, but they could not have created the basin in the first place.

While it seems likely that the Inner Basin formed as a result of collapse of the volcano's east flank, the cause of collapse remains uncertain. Flank bulging and eventual collapse caused by magma intrusion, as happened at Mount St. Helens, may or may not have been involved. An answer is probably recorded in the deposits from the collapse, but this "smoking gun" is buried by even younger volcanic deposits erupted from the host of volcanoes that dot the landscape just east of the peaks. So, although the Mount St. Helens model is attractive, mystery continues to shroud the story behind the formation of the Inner Basin.

If the Mount St. Helens model is applicable to San Francisco Mountain, the formation of the Inner Basin may well have been the single most energetic and violent eruption of the entire volcanic area around Flagstaff. And since eruptive power is generally a direct function of the volume of erupted materials, the formation of the Inner Basin may have been even more energetic than the Mount St. Helens blast.

LAVA DOMES

A *lava dome*, as the name implies, consists mostly of lava, rather than cinders or other loose volcanic fragments. Because formation of a dome-shaped pile of lava requires relatively viscous magma that can't flow far laterally, this kind of volcano typically is dacitic or rhyolitic. Domes come in sizes that range from hundreds to more than two thousand feet tall.

Lava domes also come in both monogenetic and polygenetic varieties, and because geologists have a fondness for coining many technical-sounding terms, domes are further classified as either *endogenous* or *exogenous*. An endogenous dome is one that grows entirely by injecting new magma into the *interior* of the growing dome. This growth process is somewhat analogous to inflation of a balloon, though instead of rubber stretching, the outer skin of a growing dome cools to a solid rock that breaks into jumbled and

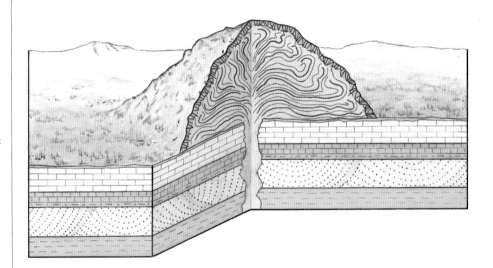

Internal structure of a typical endogenous lava dome.

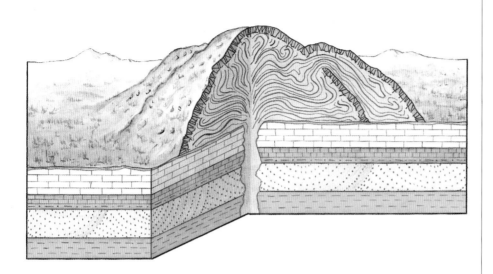

Internal structure of a typical exogenous lava dome.

tumbled rock fragments, which accumulate as cover to the still-molten and expanding interior. When eruption ends and the entire dome has cooled and solidified, the interior is a coherent mass of lava enveloped by a mantle of broken rock, called **breccia.**

An exogenous dome can be thought of as a flawed endogenous one. During exogenous formation, initial dome growth is by inflation from within (endogenous) and later growth is by magma breaking out and adding a new layer of lava, with its own breccia envelope, to the *exterior* of the early-formed endogenous body.

The characteristic internal structure of an endogenous dome is described as onion skin, because the lava interior forms concentric shells that mimic layers of an onion. The internal structure of an exogenous dome is more complex and generally includes multiple onion-skin lava layers, separated by zones of breccia.

The only sure way to correctly classify the style of dome growth is actually to see the dome form. The next best way is to see the inner parts, but this is generally impossible for domes too young to have been exposed by erosion. With experience, geologists often infer the nature of a dome's interior from the character of its outer cover. Proof of such inference must await erosional exposure or a drill core that penetrates the interior.

Sugarloaf, which sits at the entrance to the Inner Basin of the San Francisco Mountain, is a lava dome, one whose outer surface is so heavily forested that clues to inner structure are very difficult to recognize. Nonetheless, the nearly circular shape suggests a structural simplicity characteristic of endogenous domes. The somewhat corrugated aspect of the middle and lower dome flank probably is the result of erosional gullying.

In contrast, Elden Mountain, located on the east side of Flagstaff, is a structurally complex dome. The configuration of the outer surface appears to reflect an exogenous style of dome growth. Three or four separate lava lobes, each with its own particular shape, orientation and pattern of fracturing, are piled against and upon each other to create a wrinkled and corrugated aspect along the south flank of the mountain. Each lobe appears to have surfaced from fairly high on the mountain and oozed downslope to add a layer of lava to a growing exogenous body. These overlapping exogenous lava lobes are clearly visible, especially along the south flanks of the mountain.

Magma that rose through the Earth's crust to form the Elden Mountain dome pushed up some layers of sedimentary rock in trap-door fashion just before

Elden Mountain lava dome. The overlapping lobes on the south flank of the mountain are indicative of an exogenous origin.

breaking through to the surface. Had magma never actually reached the surface, geologists might have inferred the existence of a shallow but hidden dome from the presence of the upwarped and tilted sedimentary rocks. However, magma did break through to the surface, and the hidden-dome stage is recorded by east-dipping sedimentary rocks that form tilted ledges along the east flank of Elden Mountain (see Road Log, Leg 4, page 52).

The lava dome that formed within the 1980 Mount St. Helens crater went through a multi-year history of complex exogenous growth, documented by

Sugarloaf (left) is probably an endogenous lava dome, although its structure is obscured. Treeless white spots near the base of the dome are quarry pits in pumice that predates the dome. The lava dome at Mount St. Helens (right) formed over a period of a few years following the 1980 eruption. U.S. Geological Survey photos.

eyewitness accounts. Growth occurred during several periods of internal swelling punctuated by spurts of exogenous plastering of outer lava layers. More than once, the dome almost entirely destroyed itself by explosive eruptions,

but subsequent healing of blasted scars eventually resulted in the roughly hemispherical-shaped dome of today. Thus, for the Mount St. Helens dome, its simple shape belies a complicated history of exogenous growth. Might the same be true for Sugarloaf? The answer is yes, though some geologists may offer an opinion to the contrary.

Other lava domes or clusters of lava domes, most probably exogenous, in the Flagstaff area are the Dry Lake Hills, Bill Williams Mountain, Sitgreaves Mountain, Kendrick Peak and O'Leary Peak.

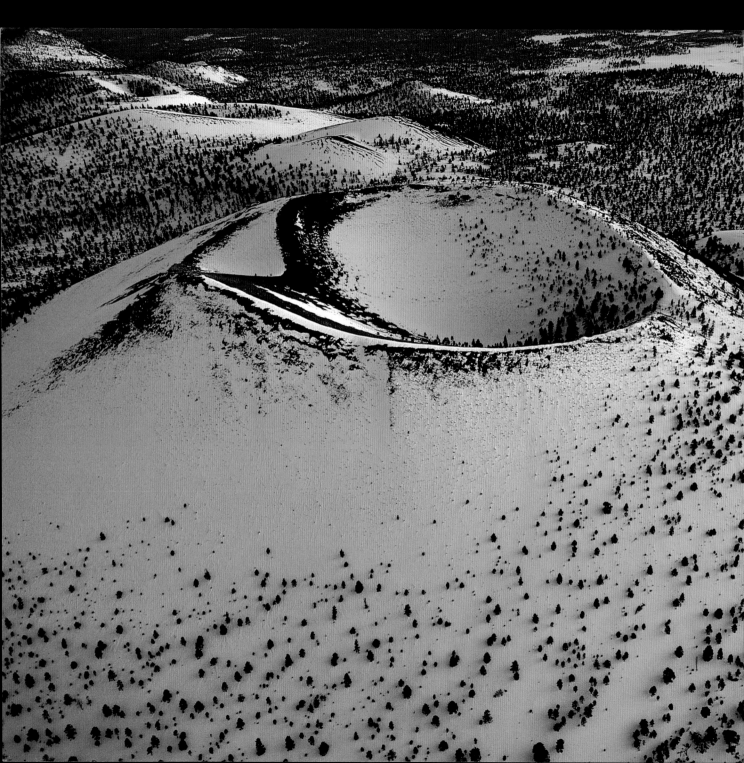

Chapter 3

The Story of Sunset Crater

The most recent eruption in the San Francisco Volcanic Field began about 930 years ago and produced the colorful cinder cone known as Sunset Crater, about 15 miles northeast of Flagstaff. Though just one of more than 600 cinder cones that decorate the landscape around Flagstaff, the uniqueness of Sunset Crater in terms of the human-family time scale is recognized by its status as a national monument.

When the Sunset Crater eruption began, the surrounding area was inhabited entirely by Native Americans. These people did not have our scientific instruments to record the earthquakes that are always precursory to volcanic eruptions. Still, the fact that no human bodies buried by Sunset Crater cinders have been discovered suggests enough understanding of what was happening to get out of harm's way. Immobile objects, however, like pit houses, were buried by cinders that

blanketed the area around the new volcano. Life near Sunset Crater must have been desirable, though, because pit houses built on top of the cinder blanket indicate the area was reinhabited not long after eruption ceased.

There is no human-recorded account of the eruption story, but the

The area surrounding Sunset Crater was occupied prior to and following the eruption in A.D. 1064. The rich prehistoric legacy is preserved at Wupatki National Monument.

annual growth rings of ponderosa pines that grew near the site where the volcano formed record the timing of the eruption about as clearly as any historian might have from first-hand observations. Deciphering the information in the growth rings is a first-class job of scientific sleuthing.

Beams of ponderosa pine were used in construction of the pit houses. Careful examination of growth rings still evident in these prehistoric beams allow scientists to precisely determine the year the tree was cut. Experts in *dendrochronology* (the study of annual growth rings of trees) tell us that the youngest ring found in beams of buried houses records the growing season of A.D. 1046. On the other hand, the oldest growth ring in beams of post-eruption houses records A.D. 1071. Thus, the eruption must have occurred after A.D. 1046 but before A.D. 1071.

Growth rings are also sensitive indicators of growing conditions. When dendrochronologists studied several trees damaged (but not killed) by the eruption, they found the rings up to and including 1064 to be normal, whereas rings for 1065 were much thinner than in preceding years. Thus, they concluded that eruption began after the 1064 growing season but before the 1065 season. A subsequent several-year period of rebound to a normal annual tree-ring pattern indicated gradual recovery from the initial eruption trauma.

A popular misconception is that the entire eruption occurred at Sunset Crater, but the story is more complex. Geologists who have carefully studied the area conclude that the eruption began with lava fountaining from a 6- to 9-mile-long, northwest-trending fissure. This style of eruption is known as a curtain of fire. After the initial fissure phase, the eruption concentrated at the northwest end of the fissure, where Sunset Crater cinder cone grew. Such an evolution, from fissure to central focus, is quite common for basaltic eruptions, and typical of the kind of magma that forms cinder cones.

Experts disagree on the duration of eruption, but available evidence suggests that phases after the initial outburst occurred intermittently for about 150 years. This evidence comes from the fact that all volcanic rocks contain small grains of magnetic minerals that are pulled into alignment with the Earth's magnetic field, just as the needle on a compass is pulled to point toward magnetic north, as magma solidifies to rock and cools. In addition, the position of the Earth's magnetic pole is known to move continuously at a rate that leaves a decipherable record of this movement in volcanic rocks if the rocks span at least 100 years. Rocks that span the entire period of the Sunset Crater eruption suggest a change in magnetic field direction, but change so near the limit of the measurement technique that skeptics can reasonably argue a contrary position.

Whatever the exact eruption duration, while inhabitants of England were fleeing attacks by William the Conqueror from Normandy in 1066, the residents of northern Arizona probably were fleeing a downpour of volcanic cinders from Sunset Crater. Local inhabitants almost certainly watched many fiery eruptions between about 1064 and 1200. Given innate human curiosity about the unusual, these shows may have been a drawing card for audiences from a broad region of the southwest.

Sunset Crater was saved from destruction in the 1920s when H.S. Colton, founder of the Museum of Northern Arizona and a prominent

Sunset Crater, on the eastern edge of the San Francisco Volcanic Field.

Flagstaff resident, thwarted the attempt of a Hollywood movie company to simulate an eruption by placing large charges of explosives in the cinder cone. Such protection is now provided by the site's national monument status. Visitors today are no longer allowed to climb Sunset Crater. This hill of loose cinders is susceptible to permanent scarring even from light pedestrian traffic. One can, however, clamber (carefully) over designated parts of the Bonito lava flow, which ponded in a small basin at the northwest foot of the cinder cone, and explore the sharp rubbly surface of the lava and several

large pieces of an early version of the cone that were rafted out on the flow like logs in a river. Waning stages of the eruption subsequently filled in areas where sectors of the early cone were rafted away, to produce the complete and nearly symmetrical cone that we see today. All of these features are so fresh and youthful in appearance that—with just a little imagination—one can visualize the eruption that created this volcanic landscape.

Given the colorful hues of the cinders that blanket the cone, the name Sunset Crater is very appropriate. However, the appearance of the volcano

can be just as intriguing at sunrise as at sunset. It seems more appropriate that Sunset Crater represent a sunrise, a beginning, rather than an end. There are sound scientific reasons to believe that Sunset Crater marks the first, but not the last, episode of historical volcanism that will decorate the landscape of northern Arizona.

THE NEXT ERUPTION

When a volcano erupting elsewhere captures national or world attention, the Arizona news media remember the home-grown variety. One of the

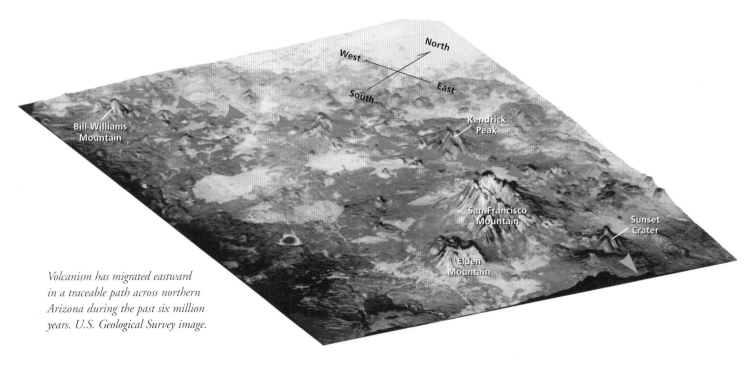

Volcanism has migrated eastward in a traceable path across northern Arizona during the past six million years. U.S. Geological Survey image.

questions inevitably asked of geologists during interviews is whether or not there will be another eruption in northern Arizona. The answer from a geologic perspective: almost certainly.

It is presumptuous for humans, whose tenure on Earth is so short, to assume that the process that produced several hundreds of eruptions spread over the past 6 million years, including one that began just 930 years ago, is now inactive. An accurate forecast of when and where the next eruption will occur is impossible. But nature has already provided enough clues for one to make an educated guess about where. And precursory earthquakes, early warnings to all volcanic eruptions, will provide a more accurate call on the where and will essentially define when.

The concepts of momentum and inertia provide help in understanding the geologic perspective on volcanism in northern Arizona. Simply restated, these basic concepts of physics say that a massive body tends to stay in a given state of motion, be it at rest or moving. Thus, it is unlikely that 6 million years of producing hundreds of eruptions within a fairly restricted part of northern Arizona has run its course with the formation of Sunset Crater.

To properly appreciate the geologic perspective, one must think in terms of geologic time. The Earth is about 4.6 billion years old, an age that most people (understandably) find difficult to relate to human events. Even 930 years, the period since the beginning of

the most recent eruption near Flagstaff, is an eternity compared to the typical human life-span. So, it is not surprising that most people think of the volcanic area around Flagstaff as extinct, or at least in deep dormancy. Precisely because of the vast difference in scale between geologic time and human-life time, geologists have pondered for generations over whether to classify Sunset Crater and similar volcanoes as extinct, dormant, or active.

Geologists recently attempted to help bridge the conceptual chasm between these two time frames through a detailed study of the life history of a single volcano, Mount Adams, in the Cascade Range of Washington state. These geologists remind us that what is called dormancy is strictly a human notion and therefore tied to our concept of time. However, such dormancy need not represent any fundamental change in geologic conditions associated with a volcano. Based on a very complete and thoroughly documented chronology of the many eruptions that built Mount Adams over a period of 500,000 years, these geologists persuasively argue that eruption-free periods of 30,000-year duration did not mean that the mantle supply of magma had dried up beneath that stratovolcano. Thus, what many geologists might describe as volcano dormancy probably represents no change whatsoever in the magma-generating and volcano-building process.

The bottom line: Sunset Crater can be classified as either dormant or

active, but however one chooses to think of it, eruptions in the vicinity are expected sometime in the near future—geologically speaking. Will future generations take that as a warning? It's anyone's guess. It seems to be a human trait to forget, ignore or otherwise downplay lessons learned just a generation or two (to say nothing of centuries) ago.

Whether or not you believe that another volcano will erupt near Flagstaff, it's worth knowing something about how magma announces its pending arrival and eruption. The well-documented eastward migration of active volcanism during the past 6 million years from near Williams to Sunset Crater suggests that the next eruption will be in the vicinity, or to the east, of Sunset Crater. A precise location for the next eruption is impossible to define until magma rising toward the Earth's surface triggers precursory earthquakes, as crustal rocks are cracked and shouldered aside to make room for magma to rise. Earthquakes of this sort are generally too weak to be felt, unless one happens to be directly over the source of shaking. However, the upward path of magma intrusion can be charted with *seismometers,* sensitive electronic recording instruments. Enough seismometers are now in place across northern Arizona to provide an early warning of rising magma. One such instrument is on display at the Visitor Center at Sunset Crater Volcano National Monument.

A Closer Look:
Volcanoes as an Educational Resource

The volcanic landscape of northern Arizona provides a well-equipped outdoor classroom and laboratory for students, teachers and tourists alike. The variety of volcano types, their excellent state of preservation, and generally unhindered accessibility make them valuable educational resources. Geology students and instructors come from across the nation and around the world to benefit from these resources. Tourists come by the millions annually. Campfire talks about the surrounding volcanoes are popular.

The volcanic landscape of northern Arizona filled a unique niche during the 1960s as the United States prepared for exploration of the moon. Astronauts needed geologic training, and Flagstaff's volcanic backyard was a classroom of choice.

Before human exploration, the moon was thought to consist mostly of basalt (eventually confirmed from rock samples collected by astronauts on the moon). The great majority of volcanoes around Flagstaff are basalt. Cratered and pock-marked regions of the lunar surface, evident through telescopes, as well as to the naked eye, were partly recreated a few miles east of Flagstaff in cinder deposits near Sunset Crater.

Dynamite charges of various sizes were positioned and exploded in a sequence appropriate to reproduce relative size, and even relative age for overlapping craters.

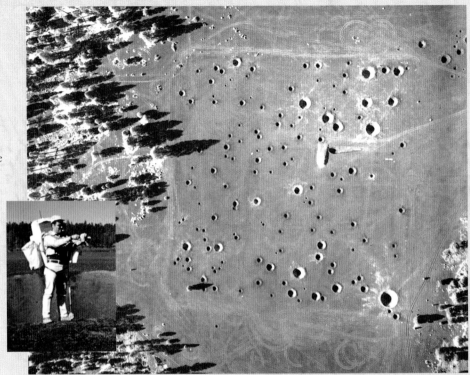

This simulated moonscape, about a mile south of Sunset Crater, was created by the U.S. Geological Survey prior to the first manned space flights to the moon, to represent the terrain astronauts might encounter there. The ponderosa pines casting long shadows are close to 100 feet tall. U.S. Geological Survey photos.

Then, in preparation for the real thing, astronauts in full lunar attire walked this pseudo lunarscape and test drove early versions of the lunar rover through the rugged cratered ground. This is the same general area where Native Americans probably once stood to watch eruptions of Sunset Crater. Today, any sense of history, training, exploration and discovery associated with this ground is lost to the sound of off-road vehicles, whose drivers find the craters an irresistible temptation.

If many earthquakes originate at successively shallower depths beneath a small geographic area, this suggests the ascent of magma beneath that area. In addition to many discrete small earthquakes, magma pulsing upward through the Earth's crust commonly causes steady earth shaking, called volcanic or *harmonic tremor*, that can last minutes to hours. When they occur together, upward-migrating earthquakes and harmonic tremor are strong evidence that magma is on the rise.

Based on studies of volcanic areas around the world, the time between onset of such earthquakes and eruption is likely to be on the order of days, weeks or even months. Lest this seem like frighteningly little advance warning for us to adequately react to the threat of a volcanic eruption, remember that the possibility of *any* eruption near Flagstaff during our lifetime is too small to lose sleep over. Meanwhile, infrequent and isolated earthquakes in northern Arizona record not magma movement but normal restlessness of the Earth's crust.

CLOSING THOUGHTS

Whether volcano lovers like it or not, northern Arizona will always be better known for the Grand Canyon than for the volcanic hills seen along the roadways leading to the canyon. Nonetheless, the mindful naturalist realizes that the combination of these two neighboring geologic features provides an even stronger lesson in Earth history than either one alone. Synergism is alive and well in the northern Arizona landscape.

The closing lesson, which carries a bit of irony with it, is that the Grand Canyon and its nearby volcanic neighbors originated simultaneously, or nearly so, through diametrically opposed processes. The blank canvas upon which this history was written was the same: an extensive plateau underlain by thousands of feet of nearly horizontal layers of sedimentary rock. But as the process of erosion was cutting downward into the pre-canyon plateau over the past 6 million years or so, volcanic processes were building upward from this sedimentary base. The resulting picture is one of great contrasts—about 10,000 feet worth, from the bottom of the Grand Canyon to the summit of San Francisco Mountain.

The Grand Canyon is fundamentally the product of destructive forces. Time and water have worked together to dissolve and disaggregate layers of rock strata, moving about 1,000 cubic miles of these pirated materials downstream to the Imperial Valley and the Gulf of California, where the Colorado River comes to its end. Bits and pieces from rock formations exposed in the walls of the canyon are now the framework of some of the world's most productive agricultural soils, the delta of the Colorado River. Limestone to lettuce! Sandstone to salad!

Meanwhile, before serious canyon cutting got underway, volcanoes began to bury part of the plateau with dark-colored basaltic lava flows—chocolate frosting on a colorful sedimentary layer cake. Nearly 120 cubic miles of lava spread over an area of 1,800 square miles. The center of the "cake" was decorated with a chocolate-chip-shaped volcano, San Francisco Mountain, the highest point in Arizona. Today, the partly nibbled tip of this chip rises as far above the plateau as the canyon cuts into it.

This story of opposing forces of nature is incomplete. The Colorado River continues its program of destruction on a grand scale, while Vulcan's massive land-construction project next door goes forward at an uneven pace. It's a living example of geology's rock cycle. Rock is removed. Rock is added. Over geologic time, the net change is zero. For those of us fortunate enough to be present at this instant in geologic time, it is a gift, part of the glory of this landscape we call northern Arizona.

Chapter 4
Road Logs

Within the San Francisco Volcanic Field, a few geologic features dominate the landscape, appearing again and again as one crisscrosses the area by auto. Foremost is the cinder cone, several hundred of which give rise to the characteristically hilly terrain of the area. These cones range from a few hundred to over a thousand feet tall. Some are tree-covered while others are nearly barren, depending mostly on local variations in climate. All had a common fiery origin: fountains of basaltic lava built cone-shaped piles of cinders around the base of the fountain.

Lava flows are at least as common as cinder cones, though they aren't so noticeable because they don't stand high above their surroundings. Every eruption that produced a cinder cone also produced lava flows. Flow colors range from black, to rusty red, to a rather pale gray. Most flows have rubbly surfaces of tumbled broken pieces of rock. Flow interiors, though, are commonly dense, broken only by partings as rock cooled and shrank. Flows form flat to gently dipping layers up to tens of feet thick and often a mile long or longer.

A third noteworthy feature is the generally flat and treeless area called a park. Within the San Francisco Volcanic Field, parks are grassy meadows surrounded by volcanic hills. Most are simply areas that have been excluded from volcanic activity. They are surrounded by volcanoes, but were themselves spared from a fiery burial. Parks are of special significance to humans, because (in contrast to the steep slopes of volcanoes) they include many sites convenient for home construction, and they commonly contain shallow deposits of groundwater, an extremely valuable resource in northern Arizona. Silt, sand, and gravel that wash into parks from the flanks of surrounding volcanoes form a local accumulation of sediment that is favorable for holding groundwater. Water wells drilled into such pockets of sediment are often only 100 to 200 feet deep, whereas wells that tap into the regional ground water table that underlies all of northern Arizona typically must be between one thousand to two thousand feet deep. Because ground water in parks sits high above the regional water table, it is described as "perched."

Finally, watch for quarries in cinder cones. These are obvious manmade scars whose walls provide a view of the internal structure of a cinder cone. With rare exception, the cinder cones of the San Francisco Volcanic Field are otherwise too young for erosion to have sculpted cutaway interior views.

USING THE ROAD LOGS

The following road logs take you on self-guided tours of volcanic features that can be seen, and in some cases walked to, from the main roads that cross the San Francisco Volcanic Field. Each log has an easily identified starting point. Features of interest are located by distance from the start, as recorded

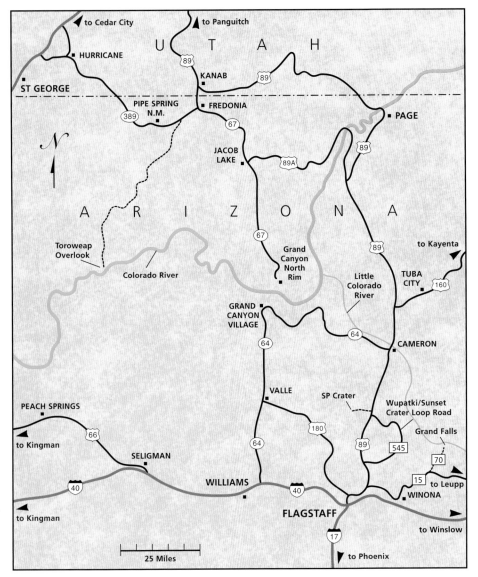

by your vehicle's odometer. Because all odometers are not equally calibrated, features are furthermore located, when possible, relative to mileage posts (MP) that appear as 3-feet-tall signs planted each mile along the shoulder of the road. Directions are generally given as east or west, north or south, where it is clear; when in doubt refer to the maps provided. Forest roads are designated by number; elevations noted are in feet above sea level.

Drive carefully and park well off the highway should you decide to stop for closer examination. Unpaved roads are impassable when wet. During summer months, when daily thunderstorms are the norm, that can be every afternoon—so use caution when venturing onto dirt roads.

WARNING
READ BEFORE USING ROAD LOGS

Although every effort has been made to accurately describe the road logs in this book, discrepancies between text and actual conditions may exist. Some natural features and conditions may have changed. Hazards may have increased or new ones may have appeared since publication of this book. When you follow any road log, you assume responsibility for your own safety. Such trips require careful attention to traffic, road conditions, terrain, weather, the physical limitations of your party, and other factors. Knowledge of current conditions and the exercise of good judgement are essential for a safe, enjoyable self-guided tour of the features described in the road logs.

There are no express or implied warranties that these logs contain accurate and reliable information. There are no warranties as to fitness for a particular purpose. Your use of the road logs expressly indicates your assumption of the risk that it may contain errors and is an acknowledgment of your responsibility for your own safety.

leg 1
Grand Canyon National Park to Williams via Highway 64

0.0 South Entrance Station, Grand Canyon National Park.

11.0 Junction with Forest Road 305 on the left. This dirt road (impassable when wet) circles the base of Red Butte. About 3 miles off the highway, on Forest Road 340A, there is a small parking area and a 1-mile long foot trail to the summit (see map).

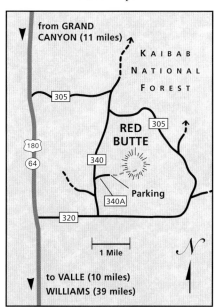

Roads in the vicinity of Red Butte.
Not all roads are shown.

13.7 (MP 224) Junction with Forest Road 320 on the left; pullout on right just beyond it. Just a stone's throw beyond the sign that indicates that you are leaving the Kaibab National Forest, Red Butte, a 1,000-foot tall reddish hill, is visible to the east, about 3 miles away. This road provides access to Forest Road 340A and the trail to the summit of Red Butte (see map).

In spite of its classic volcanic shape and 30-foot cap of basaltic lava, Red Butte is not a volcano. Beneath the lava cap, Red Butte consists mostly of reddish shale, siltstone, and sandstone of the Moenkopi Formation. At the base of the hill is Kaibab Limestone, the same formation that forms the rim of the Grand Canyon. The sedimentary rocks above the limestone were once far more extensive but have been eroded away. The cap of basaltic lava shields these relatively soft sediments from further erosion. It's an instructive example that lava is much more resistant to erosion than most sedimentary rocks.

Where is the vent from which this basalt was erupted? Geologists don't know. But at 9 million years old, the basalt itself is simply an erosional remnant of what must have been a more extensive flow.

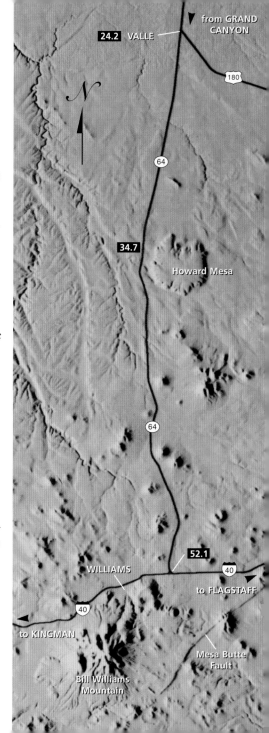

Red Butte, an aerial view of the summit.

17.0   As you continue south and emerge from the forest onto flat grassy terrain, you can see miles to the south and southeast the hilly country that is the San Francisco Volcanic Field. Each hill is a volcano, or a cluster of overlapping volcanoes. The highway thus far is built on Kaibab Limestone.

24.2   Intersection of Highways 64 and 180, at Valle. Road log continues south on Highway 64 towards Williams.

31.0 (**between MP 207 and 206**)   At this point the road climbs from Kaibab

Limestone up onto basalt, at the edge of the San Francisco Volcanic Field. You can see lava flows exposed in the road cut. These flows probably originated at the vent marked by a 300-feet-tall hill of cinders visible about a mile on your left (to the east). The white coating on many surfaces of the dark-colored basalt is *caliche* (pronounced ka-LEE-chay). Caliche is a thin deposit of salts left behind when water evaporates from a wet surface. You can usually scratch caliche off with a finger nail or knife blade to expose the dark-colored basalt underneath. One of the main salts is calcite (calcium carbonate), the same mineral that is the fundamental building block of limestone.

33.6   For about the next 6 miles, the road repeatedly crosses over from dark basaltic lava flows to the underlying (lighter-colored) Kaibab Limestone.

34.7 (**MP 203**)   The broad, mesa-like hill that rises 300 feet on your left (east) is a 3-mile wide, nearly circular andesitic lava flow called Howard Mesa. This circular feature is readily evident on the shaded relief map of the San Francisco Volcanic Field (pages 2-3, and 41).

The lava is about 2 million years old, and sits on top of (and therefore is younger than) the lavas exposed in road cuts. The vent is thought to be buried beneath the eastern part of the lava flow itself. Because of its andesitic composition, the lava was sticky and viscous enough to form a relatively thick mass rather than flow out in a thinner film typical of basaltic flows.

36.1   The quarry on your right (to the west) is in a basaltic lava flow. The dense lava is crushed into walnut-sized pieces that commonly are used in road construction. Cinders are also used in road construction, but being full of air bubbles (vesicles) are pretty easily ground down to powder by the weight of passing vehicles and are therefore short-lived, relative to crushed lava.

37.1   The road cut on the left (east) exposes light-colored Kaibab Limestone beneath a dark-colored basaltic lava flow. The Kaibab Limestone was

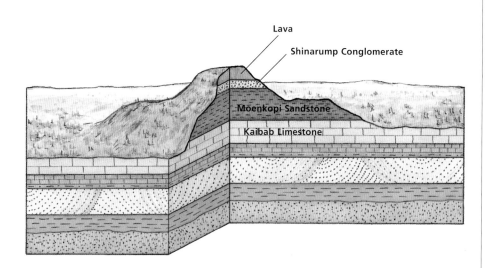

Internal structure of Red Butte.

deposited on a sea floor about 250 million years ago, whereas the basalt above was erupted on land only about 5 million years ago. The surface of contact between the two rock types is what geologists call an *unconformity* and represents a period of time for which there is no rock record. The missing record here represents about 245 million years. There wasn't always such a long gap in the rock record here, though. For example, the sedimentary rock formations above the Kaibab Limestone at Red Butte (see notes at mile 13.7) once extended this far, but were eroded away before the lava flow was emplaced.

38.7 (MP 199)   Just ahead, the road climbs a 30-40 foot hill of basalt and stays on basalt all the way to the intersection with Interstate 40, near Williams. You are now firmly within the San Francisco Volcanic Field.

44.6 (MP 193)   The large, irregularly shaped mountain about 10 miles straight ahead on the skyline is Bill Williams Mountain. It is easily distinguished from the surrounding smaller and more symmetrical cinder cones. This mountain is a composite of several lava domes and flows that built against, and on top of, one another from many closely spaced eruptive vents between 4.2 and 2.8 million years ago. Most of the lava was relatively sticky and viscous and thus tended to pile up rather than flow outward. The top of the pile (the summit of Bill Williams Mountain) is 9,300 feet above sea level.

46.5 (Spring Valley Road, just short of MP 191; pullout on left) The quarry behind you on the left exposes the innards of a basaltic cinder cone. Note that cinders in some parts of the exposure are black, but red in other parts. This two-tone arrangement of colors is true for many cinder cones worldwide. The fresh, unaltered color is black, and the modified color is red. This color change normally occurs when hot water or steam percolates through a new cone as it cools, oxidizing dark iron-bearing minerals to a red mineral called hematite. Presumably, the parts of a cinder cone that remain black did not serve as pathways for water and steam.

52.1   Intersection of Highway 64 and Interstate 40.

END

0.0 The starting point is the inter-section of Interstate 40 and Highway 64. Interstate 40 wends eastward through forested, round-topped hills, most of which are cinder cones. It also crosses a few parks, the grassy meadows created when volcanoes grow around, but not in, an area.

4.2 (MP 170) The large, asymmetrical, lumpy-topped mountain obvious about 7 miles on the left (to the north)

is Sitgreaves Mountain. Like Bill Williams Mountain, Sitgreaves is a complex pile of lava domes and flows of viscous lava erupted from several closely spaced vents. The mountain grew between 2.8 and 1.9 million years ago. Bill Williams Mountain, Sitgreaves Mountain, and Kendrick Peak (a few miles northeast of Sitgreaves) follow a northeast line along a major fracture in the Earth's crust called the Mesa Butte Fault. Magma that erupted to build

these volcanic mountains probably made its way from some depth below the Earth's crust along this fault. The fault itself is not visible along the highway but is readily evident on the shaded relief map (below).

8.4 Outcrops on both sides of the road expose thick basaltic lava flows.

10.8 (between MP 176 and 177) In the distance, straight ahead, is San

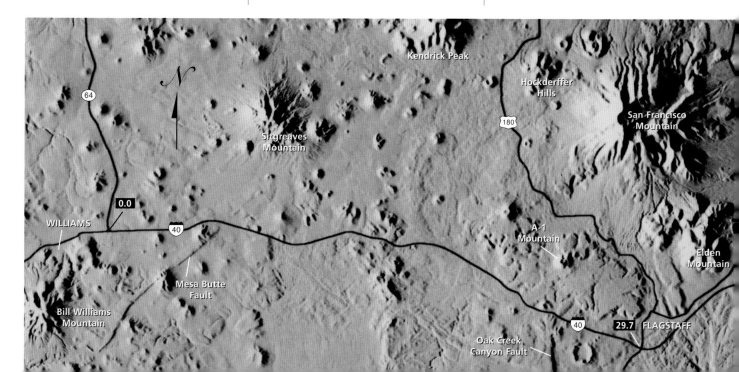

Francisco Mountain, the one and only stratovolcano of the San Francisco Volcanic Field. The now-missing top of this volcano was about 3,400 feet higher than the present summit, Humphreys Peak.

12.9   Another basaltic lava flow is exposed in road cuts on both sides of the road.

15.1 **(just short of MP 181)**   The road cut to the left exposes a lava flow thicker than is typical of basalt. This 1-million-year-old flow's chemical composition is intermediate between basalt and rhyolite (see page 9), and was therefore sticky and viscous enough to pile up in a thick mass during eruption, rather than flow out in a thin sheet.

18.0   For the next 3 miles, Interstate 40 crosses a large park.

22.6   Road cuts visible just ahead on the left are thick lava flows from A-1 Mountain, a cinder cone 2 miles to the northeast. These flows were viscous at the time of eruption (about 300,000 years ago) and are therefore thick.

25.1   At the far end of the road cut on the left, the interstate crosses the trace of the Oak Creek Canyon fault. Displacement here along this fault is a few hundred feet, down to the east, which accounts for the abrupt eastern edge of the road cut. Oak Creek Canyon, through which Highway 89A passes to the south en route to Sedona, is developed along the trace of this fault.

28.6 **(just west of MP 195)**   The road here begins its descent eastward 250 vertical feet off basalt, out of the San Francisco Volcanic Field, onto the underlying Kaibab Limestone.

29.7   Intersection of Interstates 40 and 17 at Exit 195 B. Kaibab Limestone is well exposed in outcrops on the exit ramp.

END

Leg 3
Flagstaff to Valle
via Highway 180

0.0 The starting point is at the Museum of Northern Arizona in Flagstaff. The museum itself is built on a 6-million-year old basaltic lava flow. Just north of the museum on Fort Valley Road (Highway 180) the large stratovolcano, San Francisco Mountain, looms ahead, a bit to the right.

1.1 The roadway narrows and passes through a grove of aspen trees. The road here follows the path eroded by the Rio de Flag 20 to 30 feet down into an andesitic lava flow from San Francisco Mountain. The road then climbs onto and across this lava flow for the next mile or so.

3.4 (just before MP 222) A turnoff on the right, with parking, provides a good opportunity to view Agassiz Peak (12,356 feet elevation), the highest point visible from here on San Francisco Mountain, directly to the north.

4.5 (MP 223) You are now entering Fort Valley, a park or meadow surrounded by volcanoes. The most prominent surrounding volcanoes are A-1 Mountain cinder cone to the south, Wing Mountain cinder cone to the west, and San Francisco Mountain

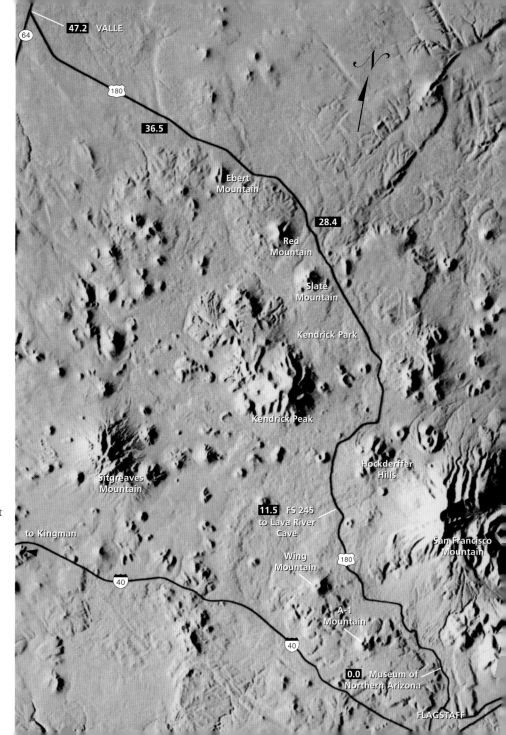

64

47.2 VALLE

180

36.5

Ebert
Mountain

28.4

Red
Mountain

Slate
Mountain

Kendrick Park

Kendrick Peak

Hockderffer
Hills

Sitgreaves
Mountain

11.5 FS 245
to Lava River
Cave

to Kingman

Wing
Mountain

180

San Francisco
Mountain

40

A-1
Mountain

40

0.0 Museum of
Northern Arizona

FLAGSTAFF

stratovolcano to the north. The 100 to 200-foot steep-sided ledges that bound the valley on the south and west are the lateral edges of lava flows from the A-1 and Wing Mountain cones. The Wing Mountain lava flow is about 1.3 million years old, and the A-1 flow about 300,000 years old. The A-1 flow spread as far as 4 miles to the east and 2 miles to the south from its cinder-cone vent. The edge of the flow, well preserved because of its geologic youth, forms the steep hill west of the Museum of Northern Arizona and all along the west side of Flagstaff. Lowell Observatory is built on top of this flow. The south

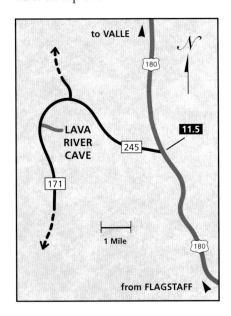

Roads in the vicinity of Lava River Cave.

edge of the flow is exposed in road cuts at mile 22.6 of the Williams to Flagstaff road log.

5.5   As the road leaves Fort Valley, it climbs gently for about 4 miles onto lava flows that form the south and west flank of a basaltic shield volcano (informally called the Hart Prairie shield) whose center is about 2 miles east of the highway, on the west-facing flank of San Francisco Mountain. This shield volcano was built about 700,000 years ago, and some of its lava flows extend more than 10 miles to the west.

A well-known lava tube exists in flows of the shield, west of the highway. To visit this tube (known as Lava River Cave), turn left (west) on Forest Road 245, 6 miles ahead, just beyond Mile Post 230.

10.2 (just before MP 229)   The road enters a stand of large aspen trees. The tops of the Hockderffer Hills are sporadically visible about 3 miles straight ahead. These are a cluster of rhyolitic lava domes (piles of thick viscous lava) and basaltic cinder cones.

11.5 (Forest Road 245)   This is the road to Lava River Cave. Turn left and Drive 3 miles to a "T" intersection with Forest Road 171. Turn left (south) and drive about 1 mile to an intersection with Forest Road 171B. Turn left and drive 0.5 mile to a foot trail that leads to the skylight cave entrance about 300 feet away. **Please remember**

that these unpaved roads are impassable when wet.

13.6   Behind the Nordic Ski Center, on the east side of the highway, are the Hockderffer Hills again, with a good view of the west side of San Francisco Mountain in the background. The peaks are Agassiz (12,356 feet) on the right, and Humphreys (12,633 feet), the highest point in Arizona, on the left.

14.4 (MP 233)   The large mountain about 4 miles on your left (to the west) is Kendrick Peak. This volcanic mountain rises to a height of 10,418 feet above sea level and consists of many lava flows and lava domes built against and on each other between 2.7 and 1.4 million years ago.

15.1   Elevation: 8046 feet. This is the highest point along Highway 180 as it crosses the San Francisco Volcanic Field.

17.0   The road enters Kendrick Park, a meadow or park that formed in the same manner as Fort Valley. To the south is an excellent view of the northwest slope of San Francisco Mountain, and of Humphreys Peak (12,633 feet).

18.8   The road leaves Kendrick Park and descends 200 to 300 feet through a series of basaltic lava flows, visible in outcrop, until it flattens out over a roughly horizontal surface of lava for the next few miles.

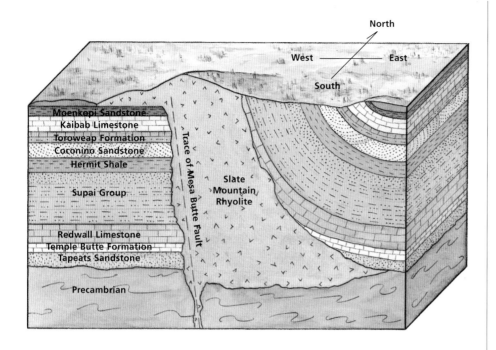

Moenkopi Sandstone
Kaibab Limestone
Toroweap Formation
Coconino Sandstone
Hermit Shale
Supai Group
Redwall Limestone
Temple Butte Formation
Tapeats Sandstone
Precambrian

Trace of Mesa Butte Fault

Slate Mountain Rhyolite

North
West
East
South

Slate Mountain. The intrusion of magma which created Slate Mountain, a lava dome, has pushed up the sedimentary layers on its east side.

24.8 The lumpy mountain visible to the left is Slate Mountain, a lava dome. Slate is the name for a dark-colored metamorphic rock that breaks into large thin sheets, which once were widely used as blackboards and shingles. But there is no slate in Slate Mountain. Instead, the lava simply breaks into slate-like sheets, thus giving the mountain its name. The lighter-colored treeless hill in front of Slate Mountain is formed by Permian and Triassic sedimentary rocks that were pushed up and tilted by magma that rose to form the rhyolite dome of Slate Mountain about 1.5 million years ago. Kendrick Peak is visible to the south.

28.4 (MP 247) A sign on the left announces the road to the Red Mountain Geologic Area. There is a parking area just off the highway, and **a dirt road (impassable when wet) continues 0.3 mile beyond to a trailhead. A gentle** uphill walk (allow about 40 minutes each way) leads to an amphitheater-like bowl eroded in cinders on the flank of Red Mountain. Note that all the cinder layers are inclined north to northeast, instead of dipping radially away from a central point, as is generally the case for cinders that accumulate around a vent in calm weather. This configuration is probably the result of strong winds from the southwest at the time of eruption. Approximate date for the eruption of Red Mountain is 740,000 years ago.

The cinders adhere to each other in outcrops, rather than forming a pile of loose pieces. The individual cinders probably are cemented together by thin mineral coatings deposited by water or steam that percolated through the cinders shortly after they accumulated.

You can see pea-sized pieces of a black glassy-looking material both within cinders and as loose grains, especially as sand in the bed of the small stream that drains from the amphitheater. They are crystals of a mineral called hornblende, though they look much like the volcanic glass called obsidian. Less abundant, but perhaps more conspicuous because of their larger size, are white-to-transparent inch-wide crystals (in some cinders) of the mineral called feldspar.

Back on the highway, you are driving over the featureless, flat surface of basaltic lava flows. The reddish color is typical of weathered basalt due to the oxidation of iron in the rock.

Red Mountain. The layering in the cinders shows as color bands across the photo.

33.8 The reddish hill to the south is Ebert Mountain, a basaltic cinder cone with abundant oxidized fragments. The small gray treeless area is a cinder quarry.

34.0 The half dozen or so hills in the distance to the north are cinder cones along the northern edge of the San Francisco Volcanic Field.

The flat areas beyond the cones are horizontal sedimentary rocks of the Kaibab Limestone.

36.5 (MP 255) At this point you leave the San Francisco Volcanic Field. The change from basaltic lava to sedimentary rock is subtle, even to a trained geologist. However, the edge of the volcanic field is visible as an abrupt step in elevation on the shaded-relief image (see page 47). From here the road stays on the Kaibab Limestone all the way to Valle.

47.2 The intersection of Highways 64 and 180, Valle.

END

Leg 4
Flagstaff to Wupatki National Monument via Highway 89

0.0 The starting point is the entrance to the Elden Lookout Trail parking lot 0.2 mile east of the Peaks Ranger Station, Coconino National Forest, on Highway 89, on the east side of Flagstaff. The 3-mile-long Elden Lookout Trail ascends 2,300 feet to the top of 500,000-year-old Elden Mountain, a lava dome whose steep sides reflect the high viscosity of the dacitic lava that built the mountain.

0.2 The cinder cone nearest the highway on the right is Sheep Hill, extensively quarried for cinders. Parts of other quarried cinder cones are visible to the right (south) of Sheep Hill.

1.3 Traffic signal at the intersection with Townsend/Winona road. This is the turnoff for Grand Falls (Road Log, Leg 7, page 59). Continue north on Highway 89.

2.1 Just behind you to the left, the first hill on the skyline is Elden Mountain. The low, tree-covered hills at the base are much older sedimentary rocks that were pushed up and tilted by the magma that rose to the surface to form Elden Mountain (see illustration on page 52).

4.0 Straight ahead, the tallest volcano on the horizon, emblazoned with the zigzag scar of road cuts and topped by twin peaks, is 250,000-year-old O'Leary Peak, a steep-sided lava dome built of the same kind of sticky viscous lava (dacite) that formed Elden Mountain.

5.0 To the left is a good view of San Francisco Mountain stratovolcano. Note Sugarloaf, on the flank of San Francisco Mountain. Sugarloaf is a rhyolitic lava dome. To the right, many round-topped, partly forested and barren basaltic cinder cones dot the landscape.

9.3 Road cut on the right exposes part of a cinder cone.

10.0 Road cut on the left exposes the interior of an andesitic lava flow erupted from San Francisco Mountain stratovolcano.

11.1 Turnoff (on the right) to Sunset Crater Volcano National Monument. Continue north on Highway 89.

11.8 To the left is a quarry that exposes the interior of a cinder cone. **The dirt road to the left is Forest Road**

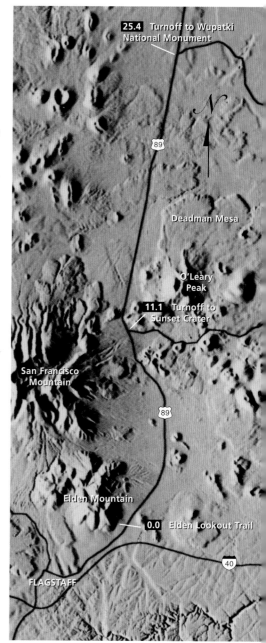

552 (impassable when wet) which leads to the Inner Basin of San Francisco Mountain, where the interior of this stratovolcano is exposed. For the next 3 miles, Highway 89 descends over a thousand feet in elevation.

13.5 Many basaltic cinder cones appear as round-topped hills to the left.

16.0 About a mile to the east, the abrupt 300-foot step up onto a relatively flat-topped hill is the edge of a 170,000-year-old dacitic lava flow called Deadman Mesa. The highway continues over the tops of basaltic lava flows.

20.9 (just beyond MP 440) Pullout on right. Light-colored hills, 10 to 20 feet tall, of disturbed ground on both sides of the highway mark an area where pumice was once mined. Pumice is the rhyolitic equivalent of cinders, which are basaltic. This pumice was used as an additive, called pozzolan, to increase the strength of concrete used in the construction of the Glen Canyon Dam at Page, Arizona. If you stop and examine the mined ground, you will find remnants of pumice deposits that form blanketing layers over a pre-existing hilly terrain. The layering in the pumice mimics the shape of the landscape on which the pumice fell. This pumice was erupted explosively about 800,000 years ago, probably from a vent on San Francisco Mountain.

21.4 The road cuts through a basaltic lava flow on the left, younger than the pumice, and continues over other lava flows.

24.0 Eroded edge of basaltic lava flow on the left.

25.4 (between MP 444 and 445) Turnoff to Wupatki National Monument.

END

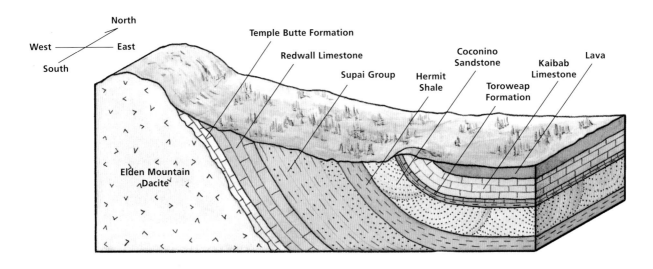

Elden Mountain. The intrusion of magma which created Elden Mountain, a lava dome, has pushed up sedimentary layers on its east side.

Leg 5
Wupatki/Sunset Crater Loop from Highway 89

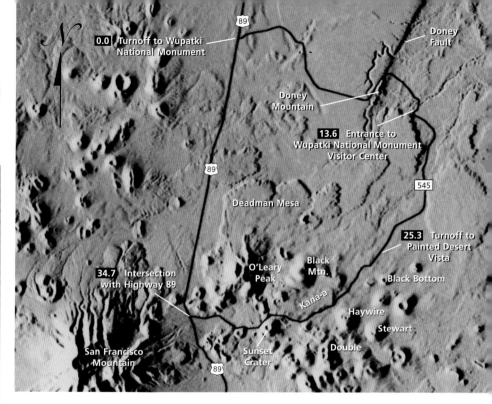

0.0 The starting point is the pullout at the Wupatki National Monument sign, at the north end of the Wupatki/ Sunset Crater Loop Road (Forest Road 545), just east of Highway 89, 25 miles north of Flagstaff. Note that 3-foot-tall, tan-colored mileage markers are planted in the road shoulder at 1-mile intervals for the next 34 miles, at which point the road intersects Highway 89 to the south. For the first 0.7 mile, the road meanders across a grass-covered surface of basaltic lava flows. The colorful landscape in the far distance, miles to the east, is the Painted Desert, in which multi-colored sedimentary rocks of Mesozoic age (the Chinle Formation) are exposed.

0.7 The road descends off the edge of a basaltic lava flow and onto the Kaibab Limestone, the sedimentary rock that forms the rim of the Grand Canyon. This light-colored limestone is well exposed in outcrops for the next couple of miles.

3.8 (at the Lomaki Ruin turnoff)
All around you, basaltic lava flows appear as black caps on the hills of Kaibab Limestone. For the next several miles, the pervasive black color of the ground is caused by a veneer of basaltic cinders barely covering the underlying Kaibab Limestone.

6.0 To the right, in the distance, notice the V-shaped notch in the side of San Francisco Mountain, opening to the Inner Basin.

9.0 The reddish hills just ahead are a group of small cinder cones aligned along the north/northeast-trending Doney fault. The tallest of these is Doney Mountain. Movement across the fault has dropped rocks on the east side down about 200 feet relative to rocks on the west side. The offset along this fault is preserved in present day topography (clearly visible in the photo on page 54).

9.4 Turnoff to the Doney picnic area. A foot trail ascends some of the cinder cones from the picnic area. From the top (or from the parking lot), on a clear day you can see flat-topped, tower-like landforms about 40 miles away on the eastern horizon. These are the Hopi Buttes, a group of volcanoes that erupted 7 to 8 million years ago— just before those of the San Francisco Volcanic Field. Unlike those of the San Francisco Volcanic Field, eruptions that formed the Hopi Buttes occurred in

View to the south along Doney Fault. The vertical offset along this fault (approximately 150 feet) is clearly evident in the escarpment visible here. San Francisco Mountain is visible on the horizon.

shallow water or on water-saturated mud. As a result, a typical eruption included powerful steam explosions that cored out a crater. Many such craters were filled with lava as eruption subsided, and these lava ponds hardened to erosion-resistant rocks that became the caps of flat-topped buttes once erosion removed surrounding soft material.

10.0 The road starts to descend over the escarpment of the fault. Layers of cinders on the flank of Doney Mountain are exposed in the road cut on the right, and the truncated edges of faulted limestone beds are exposed on the left. The fact that the layers of volcanic deposits are not broken and offset by the fault indicates that the fault and its escarpment were present before the volcanic eruptions.

11.2 Pullout on right. Looking back towards Doney Mountain, note the alignment of the cinder cones along the trace of the Doney fault. Magma that fed the eruptions apparently found the pre-existing fault a convenient pathway through the Earth's upper crust. To the north are the older sedimentary rocks: light-colored Kaibab Limestone and the overlying reddish Moenkopi Formation. Note that the Moenkopi has been totally eroded off the high (west) side of the fault escarpment. At the base of this escarpment, you are off the NE edge of the San Francisco Volcanic Field. The road now stays on sedimentary rocks (mostly Moenkopi Sandstone) for several miles.

13.6 Entrance to Wupatki National Monument visitor center. Most of the Indian ruins at Wupatki are constructed of the local sedimentary rocks. However, pieces of basaltic lava were also used in wall construction (see photo on page 55). The 300-foot-tall mesa immediately behind the ruins is capped with a 1-million-year-old lava flow that is a likely source of this building material.

16.8 The road crosses the contact between Kaibab Limestone and the overlying basalt and climbs onto a lava flow. Here you are re-entering the San Francisco Volcanic Field. The thin white coatings on some pieces of basalt are caliche, principally a calcium carbonate salt left behind when a wet lava surface dries.

Hopi Buttes

Detail of Citadel Ruin, Wupatki National Monument. Dark-colored rocks are basalt.

18.0 For about the next mile, note the 50-foot rise to the east. This scarp marks the edge of a lava flow that is on top of, and thus younger than, the flow the road is on. Approximate age of the younger flow is 500,000 years.

23.6 The lava on the left, though not a very distinctive landform, is the far end of the 4-mile-long basaltic flow known as Kana-a, which erupted from Sunset Crater. From here, the road follows the north edge of this flow, upstream, almost to Sunset Crater itself. Though several miles long, the flow is only about 1,000 feet wide, because it erupted into and followed a narrow stream valley, a classic example of an intracanyon flow.

25 You are now surrounded by several cinder cones.

25.3 Turnoff to Painted Desert Vista for a magnificent panoramic view.

26 Note that the blanket of cinders covering the ground thickens toward Sunset Crater, their source.

27.9 The road crosses to the south side of the Kana-a lava flow.

28.3 The road crosses back to the north side of the flow. The flow then disappears immediately beneath the blanket of cinders erupted from Sunset Crater.

28.9 Note that now the cinder deposit in the road cuts on the right is at least 6 to 10 feet thick and distinctly layered. An excellent view of Sunset Crater ahead.

29.2 Entrance to Sunset Crater Volcano National Monument. Some years ago, the name was changed from Sunset Crater National Monument to Sunset Crater Volcano National Monument in order to avoid confusion with Meteor Crater (see page 14), another well-known geologic feature of northern Arizona. Meteor Crater is not a volcanic feature but an impact structure, the result of a collision that occurred 50,000 years ago when a large meteor or asteroid struck our planet.

29.3 At the crest of the hill you get an excellent view straight ahead into the Inner Basin of San Francisco Mountain.

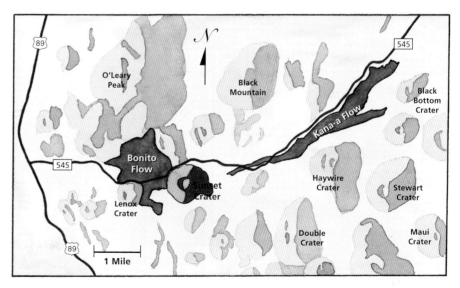

Map of Sunset Crater and vicinity, showing Bonito and Kana-a lava flows.

Sunset Crater

30.0 On the left is the steep north-facing flank of Sunset Crater. On the right is the south-facing flank of a smaller volcano, older than Sunset Crater and thus blanketed with cinders from the Sunset Crater eruption.

30.5 The road crosses rough and rubbly lava from Sunset Crater.

30.9 To the right, two dark reddish hills protrude about 50 feet above the surface of the black lava flow. These hills were once part of the Sunset Crater

cinder cone, but then broke away and were rafted out on the lava flow like logs floating in a stream. Continuing eruption filled in the void left by the missing pieces to produce the smoothly shaped cone of today.

31.1 Turnout on left for a mile-long foot trail over some of the products of the Sunset Crater eruption.

31.3 A turnout on the right gives parking and access to a foot trail out onto the lava flow from Sunset Crater,

called the Bonito Flow. Though basaltic and therefore of low viscosity, the Bonito Flow is unusually thick, because it ponded in a basin at the foot of Sunset Crater. Most of the lava surface is rough and broken into tumbled pieces that form what is called *aa* lava. Aa lava forms because the brittle outer surface of the flow breaks as the molten interior continues to move. Where the lava did not advance to the aa stage, the surface is smoother and sometimes ropy, or corrugated, in what is called *pahoehoe* lava.

32.7 Visitor Center on left.

33.1 To the west (straight ahead) is an excellent view into the Inner Basin, the missing part of San Francisco Mountain stratovolcano. The light patches near the mouth of the basin are quarries for pumice. The nearly sym-metrical, tree-covered hill behind the quarry on the right is a 200,000-year-old lava dome called Sugarloaf. Contin-uing on, just past the Sunset Crater visitor center, the road skirts along the north side of Bonito Park, a valley that formed when volcanoes grew around its periphery.

34.7 Intersection with Highway 89.

END

Leg 6
SP Crater

Cinder cones grow when basaltic or andesitic magma erupts in fiery orange and red fountains of lava spray that fall back to Earth, building a cone-shaped or round-topped hill around the erupting vent. Each individual cinder is simply a lava droplet or blob that solidifies before falling back to Earth. A lava flow may also form

A two-foot long volcanic bomb on the rim of SP Crater.

when magma of low volatile content quietly oozes from a vent, without fountaining. SP Crater, located about 26 miles north of Flagstaff, is a nearly perfect example of a cinder cone and its associated lava flow. The road to SP Crater will provide opportunities for a close look at both a classic cinder cone and its associated flow.

0.0 The starting point is the intersection of Highway 89 with the turnoff to Wupatki National Monument. Continue north from this intersection on Highway 89. SP Crater is on Babbitt Ranch land, and has been for over 100 years. It was, in fact, named by C. J. Babbitt in the 1880s. It is with their gracious permission that we include it in this log.

1.2 (**MP 446**) Hanks Trading Post on the left.

3.0 **Turn left onto an unmarked dirt road. This road is impassable when wet.**

4.0 Ahead of you, many cinder cones are visible, including SP Crater, which is difficult (from this angle) to distinguish. There are intermittent patches of lava here as well.

5.2 The andesitic lava flow associated with SP Crater is clearly visible directly in front of you as a dark ridge, about 60 feet tall, just a couple of miles ahead. Continue straight.

6.9 The road arcs gently to the left (southwest) and approaches the eastern edge of the flow.

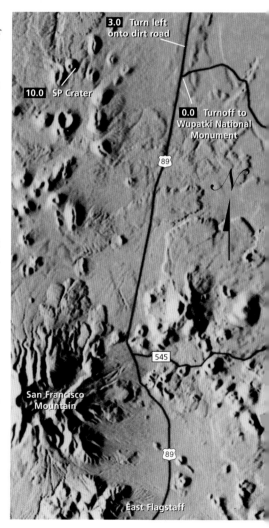

Aerial view of SP Crater and its associated lava flow.

flow is below its margins, because lava in this central channel flowed out leaving lava banks, or levees, when the supply of new lava was shut off at the vent. On its east side, the flow laps against and onto older basaltic lava flows. On its west side, it covers Kaibab Limestone. Two flow fingers protrude westward where they spilled into and partly filled a down-faulted trench (or **graben**) 1000 feet wide and 30 feet deep which was present before eruption began.

10.2   Here you are right in the middle of the lava channel created during the waning stages of the eruption, as lava flowed north, leaving the empty channel behind.

10.4   You are now on the far side of the SP flow. Turn around and return to Highway 89.

END

7.7   The road skirts the eastern edge of the flow. The blocky surface texture of the flow is apparent from the road, This texture is the andesitic equivalent of the basaltic *aa* flows.

9.0   SP Crater is clearly visible straight ahead.

10.0   Here the road begins to cross the flow. The cinder cone, immediately south of the road, is about 1000 feet tall. This cone and its lava flow are about 70,000 years old. For the physically fit, a scramble over loose cinders to the rim is rewarding. The rim consists of lava blobs and bombs up to 3 feet long that adhere to each other to form a hard rock resistant to erosion. The core of the cone is a 350-foot deep crater, from whose depths enchanting views of night skies are afforded on moonless nights.

The SP lava flow extends 4 miles to the north. The surface upon which this flow erupted tilted gently to the north (as it does today), which accounts for the direction of the flow. Outward for about 0.25 mile from the base of the cinder cone, the central part of the

Leg 7
Grand Falls

The starting point is the entrance to the Elden Lookout Trail parking lot 0.2 mile east of the Peaks Ranger Station, Coconino National Forest, on Highway 89, on the east side of Flagstaff, the same as that for Leg 4. **The one-way distance to Grand Falls is 33 miles. Of this 33 miles, 24 miles are paved, 9 miles are unpaved. The unpaved portion can be particularly treacherous in wet weather.** Much of the trip (including Grand Falls itself) is on the Navajo Indian Reservation, which occupies 16 million acres of the Four Corners region, including much of northeastern Arizona. **The Navajo Reservation is criss-crossed with unpaved roads, many of them unsigned, most of them not on any map. Follow directions in the log carefully, and don't be afraid to backtrack if you think you've taken a wrong turn.**

The many round-topped hills along the route are basaltic cinder cones. Road cuts expose cinder deposits and basaltic lava flows.

0.0   Entrance to the Elden Lookout Trail parking lot 0.2 mile east of the Peaks Ranger Station, Coconino National Forest, on Highway 89, on

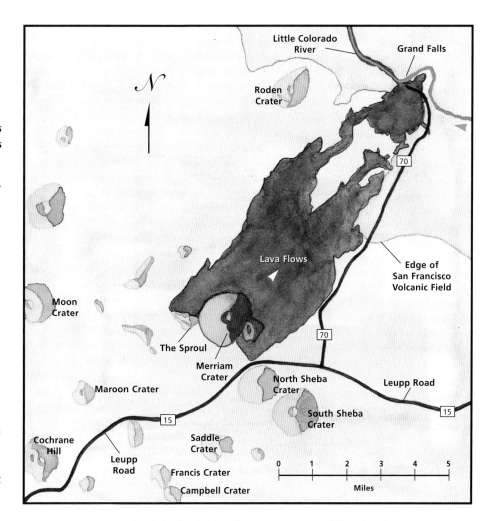

Roads in the vicinity of Merriam Crater and Grand Falls.

Aerial view of Merriam Crater, a possible source of the lava that created Grand Falls.

the east side of Flagstaff. Turn left (north) onto Highway 89.

1.3 Traffic signal. Turn right (southeast) on the Townsend/Winona Road.

9.4 Turn left (northeast) on Leupp Road (Navajo Road 15).

10.2 (**between MP 429 and 430**) Large cinder quarry in cinder cone about 1 mile to the right (east).

12.6 Road cut on left cuts through the flank of a basaltic cinder cone.

17.0 Tall round-topped hill straight ahead is Merriam Crater, a possible source of the lava flow at Grand Falls.

21.7 (**MP 441**) Merriam Crater to the left (north). Note quarry in smaller cinder cone in front of Merriam.

24.1 Cattle guard at boundary of Navajo Indian Reservation. Just a few hundred feet beyond the boundary, turn left on **Navajo Road 70. This dirt road is impassable when wet.** This road is joined by several others along the way to Grand Falls. There are many

unmarked roads and tracks on this portion of the Navajo Indian Reservation. Stay on what is clearly the largest and most heavily used road, and watch for Navajo Road 70 signs.

25.2 Over the next 0.3 mile the road descends about 100 feet off the edge of a lava flow. At the base of this hill, just beyond the roadway, is another lava flow, thinner and younger.

25.6 Grand Falls Bible Church on the left. Navajo Road 70 curves right and heads almost directly east.

27.7 Navajo Road 6920 comes in on the right.

28.4 From here on, the road is mostly on Kaibab Limestone. Dark-colored basaltic lava flows are nearby on the left (north). Hills capped with red rocks of Moenkopi Formation are visible on your right (to the south).

30.1 An abrupt 20-foot rise in the road is a fault scarp in Kaibab Limestone. This fault is indicated on the map on page 62.

31.2 Just ahead, the road crosses gravels of the Little Colorado River. Over time, the course of the Little Colorado River has migrated to the east. These river gravels were deposited by the river when its course was west of where it is today. Note the well-rounded cobbles so typical of river

*Aerial view of Grand Falls, looking upstream. The dashed line marks the path of the
Little Colorado River prior to the lava flow which altered the course of the river.*

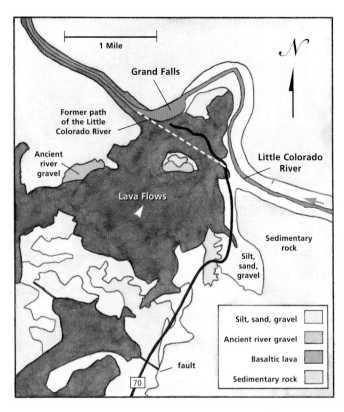

1 Mile

Grand Falls

Former path
of the Little
Colorado River

Ancient
river
gravel

Lava Flows

Little Colorado
River

Sedimentary
rock

Silt,
sand,
gravel

fault

70

Silt, sand, gravel	
Ancient river gravel	
Basaltic lava	
Sedimentary rock	

Geologic map in the vicinity of Grand Falls. Compare
this map with the photograph on page 61.

gravels. The Little Colorado River is just ahead of you.

31.5 Navajo Road 6910 enters on the right.

32.7 A left turn here will take you about 0.5 mile to a picnic area at the west side of Grand Falls. If you turn here, drive straight ahead and look for the ramadas. Park at the furthest (northernmost) ramada in the picnic area and walk another 100 feet or so downstream, to a flat promontory overlooking the river. From here you can look back into the alcove adjacent to the falls. Note the steep contact between the dark lava and the truncated horizontal layers of Kaibab Limestone in the alcove at the south end of the falls. This contact marks the east wall of the river canyon that was present when the lava flow encountered the river. If you follow upward the trace of the contact between lava and limestone, you will see that it flattens to horizontal. This change marks where lava spilled out over the east bank, or rim, of the canyon. The horizontal lava cap is only about 20 to 30 feet thick, whereas the lava that ponded in the canyon is at least as thick as the depth of the canyon below Grand Falls—about 200 feet. Note that the lava is exposed below you at the bottom of the existing canyon. The ramada where you parked is probably right above the centerline of the canyon now filled by lava. Note that the lava is broken by *joints* or fractures, some in columnar shape and others in irregular patterns. Such joints form because as lava cools it shrinks, and therefore cracks. Much of the lava is veneered with red dirt deposited by a muddy mist present when the falls is in full flow. Downstream, you can see parts of the lava attached to the walls of the river canyon and coating the river bed. The lava flowed about 15 miles downstream from here. The reservoir that was formed in the part of the canyon upstream from the lava entry has filled with sediment washed in since the lava dam formed thousands of years ago.

33.0 The Little Colorado River, about 0.2 mile above Grand Falls. **The road crosses the river and is passable (when the river is not flowing!) with a high-clearance vehicle.** If you walk down the dry river bed to the top of the falls, you can examine the horizontal contact between lava and Kaibab Limestone in the west (left) bank.

From the lip of the falls (keep safely back from the edge!) you look straight across to the lava plug that filled the canyon and thus created Grand Falls. See page 16 for more information on Grand Falls.

END

Leg 8
Toroweap Overlook

This leg begins at the junction of the Mt. Trumbull/Toroweap Overlook road with Highway 389 (see map on page 40). This junction is 8.3 miles west of the junction of 389 and 89A, **in Fredonia (your last chance to fill up on gasoline and water).** The entire trip is on **unpaved roads, through the remote portion of northern Arizona (north of the Colorado River and south of the Utah border) known as the Arizona Strip.** The route described here is by far the shortest and most accessible route to the overlook at Toroweap. There is a small campground at the end of the road (first-come, first-served) and chemical toilets, **but no water or other amenities.** It is a long, dusty drive over 60 miles of dirt road. The first 53 miles **(to the boundary of Grand Canyon National Park)** are graded and accessible to most vehicles in dry weather. The 7 miles from the park boundary to the overlook are less well maintained, can be badly rutted, and are rocky in places. All of these roads can be impassable in wet weather. The trip to Toroweap Overlook is a major undertaking under any circumstances. If in doubt about road conditions, inquire at the Bureau of Land Management (BLM) office in Kanab, Utah.

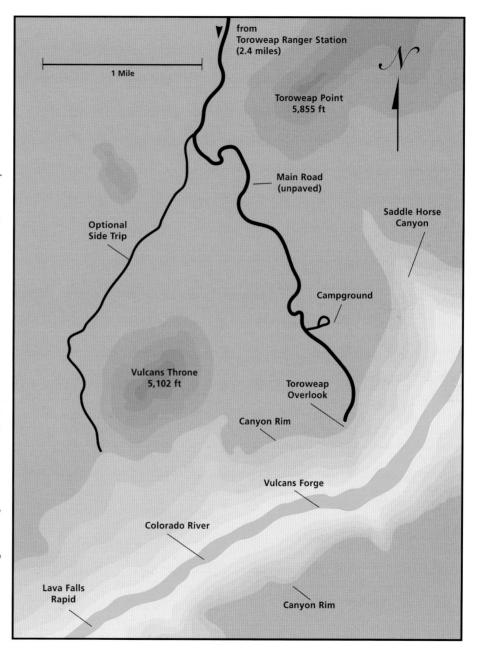

1 Mile

from Toroweap Ranger Station (2.4 miles)

Toroweap Point 5,855 ft

N

Main Road (unpaved)

Saddle Horse Canyon

Optional Side Trip

Campground

Vulcans Throne 5,102 ft

Toroweap Overlook

Canyon Rim

Vulcans Forge

Colorado River

Canyon Rim

Lava Falls Rapid

0.0   Junction of the Mt. Trumbull/ Toroweap Overlook road with Highway 389. Toroweap Overlook is 60 miles from this junction. **No water or services are available for the next 60 miles.** From here to the park boundary (53 miles ahead) you are mostly on land managed by the BLM.

18.0   Here the road descends across the scarp of the Toroweap Fault. The fault scarp will be on your left for the next few miles; offset along the fault is less than 100 vertical feet at this point. Numerous small cinder cones are visible to the southwest. On the distant horizon, Mt. Trumbull is clearly visible as the broad peak which towers above the others. Mt. Trumbull is the highest peak of the Uinkarets, a small range of geo- logically young (no older than 5 million years) volcanic mountains which form the western edge of the Toroweap Valley.

22.8   Junction with Hack Canyon road on the left. Continue straight ahead. The road is on the Kaibab Lime- stone for the next few miles, although outcrops of red Moenkopi Sandstone are visible in places.

26.0   A line of cinder cones is visible to the southwest.

27.5   Junction with side roads. Follow sign straight ahead to Toroweap.

30.1   Note the eroded edge of a basaltic lava flow on the right. This lava flow erupted from one of the southernmost of the line of cinder cones visible. These are the same cones visible at mile 26.0.

30.8   Mt. Trumbull is visible ahead, through the gap on the horizon.

36.3   The road approaches a line of basaltic cinder cones; note dark-colored outcrops of lava on the hillsides along the road.

37.5   Begin your descent into the Toroweap Valley. Mt. Trumbull is straight ahead.

39.6   The Toroweap Fault scarp is on your left and parallels the road for most of the next 20 miles, from here to Toroweap Overlook.

40.9   On your left, note that basaltic lava has spilled over the scarp of the Toroweap Fault, covering the edges of the sedimentary layers exposed in the scarp. This flow cascaded over the edge of the scarp into the Toroweap Valley and continued downstream for some distance. This is the first of several such flows which will be visible ahead (most of them on the right side of the road). For the next two miles, remnants of this intracanyon lava flow occur on both sides of the road.

42.9   You are now on the floor of the Toroweap Valley. The fault scarp rises to your left; Mt. Trumbull is to your right. The road from here to Toroweap Overlook parallels the scarp of the Toroweap Fault. Vertical offset along the Toroweap Fault is about 200 feet at this point. The scarp is capped by Kaibab Limestone.

44.0   The promontory on the horizon is Toroweap Point, which towers a thousand feet above the Toroweap Valley. The cliffs on the distant horizon are on the south side of the Grand Canyon.

46.2   Junction with the road to Mt. Trumbull on the right. Continue straight ahead.

49.5   The vertical offset along the Toroweap Fault has increased here to about 700 feet (from about 200 feet at mile 42.9). With the increased offset, one sees more of the stratigraphic section. Visible in the distance (from top down) are the Kaibab Limestone, Toroweap Formation, Coconino Sand- stone, and Hermit Shale. These are the rock units which form the upper walls of the Grand Canyon in the vicinity of Grand Canyon Village. Note, how- ever, that here, far to the west, the cross-bedded Coconino Sandstone is much thinner, and the Hermit Shale (which forms the steep red slope at the base of the cliff) is much thicker.

51.6   Looking across the valley to the west, note that the horizontally bedded sedimentary layers on the ridge are draped by lava that cascaded from the cinder cones you see along the skyline just behind that ridge. There are many examples of this on the west side of the Toroweap Valley.

51.9   Tuweep Church on the left.

53.1   Park boundary. Camping anywhere other than the campground at the end of the road requires a permit which must be obtained in advance.

53.8   Ranger station.

54.3   Vulcans Throne, the cinder cone which sits perched on the north rim of the canyon, is visible straight ahead.

57.3   The road to the right leads to the Lava Falls trailhead on the west side of Vulcans Throne. **This road is not suitable for vehicles with low clearance.** Continue straight ahead to the canyon rim.

59.1   Road to the left leads to the main campground. Sites are available here on a first-come, first-served basis, at no charge. **There are composting toilets but no water.**

60.0   Toroweap Overlook and campground. This is the end of the road.

Vulcans Throne. Photo by Nick Freedman.

Park and walk to the edge of the canyon, just a few yards to the south.

It is about 3000 vertical feet from the edge of the canyon to the Colorado River at this point; *use caution near the edge*. Volcanic features are evident all around you; note the cinder cone on the far side of the canyon directly across from the overlook, and Vulcans Throne, the large cinder cone which towers above you on the north side, just to the west.

Walk downstream from the parking area along the rim about ⅛ mile, and you will come to a promontory which offers a view (downstream) of Lava Falls rapid and the lava cascades on the north wall of the canyon. The Toroweap lava cascades, which drape the canyon wall just west (downstream) of Vulcans Throne, erupted from cinder cones further north on the western edge of the Toroweap Valley approximately 1.2 million years ago. This created a lava dam about 1400 feet high. Remnants of that dam are visible as horizontal layers of basalt at the base of the cascades.

Vulcans Throne is the prominent cinder cone perched on the north edge of the canyon. At about 700 feet tall, it is the largest cinder cone in the area; its precise age is unknown. Note that portions of the cone have slumped part way down the canyon wall. The

Toroweap lava cascades did not erupt from the same vent which produced Vulcan's Throne.

The rapid clearly visible downstream from the overlook is Lava Falls, which is formed (like most of the rapids in the Grand Canyon) by erosional debris that has washed in from tributary canyons—in this case Prospect Canyon, clearly visible on the south side of the river. Above Prospect Canyon is Prospect Valley, which extends far to the south. Both the canyon and the valley follow the trace of the Toroweap Fault as it crosses the Grand Canyon from somewhere in the vicinity of Vulcans Throne. The vertical offset along the Toroweap Fault (about 700 feet) is clearly visible: note the marked difference in elevation of the canyon rim on either side of Prospect Canyon. Note, too, the remains of a cinder cone which cling to the west wall of Prospect Canyon at the canyon rim.

The damming of the Colorado River by lava which erupted (on both sides of the river) and cascaded into the canyon occurred not once but many times during a period that lasted about 1.4 million years. At the mouth of Prospect Canyon at river level, you will see a large mass of lava. This is a remnant of one such lava dam. Near the top of Prospect Canyon remnants of the tallest such dam (the Prospect dam) are also visible; they indicate that the river was dammed to a height of over 2000 feet. The lake created by the Prospect dam (about 1.8 million years

ago—making it the oldest of the lava dams in the canyon) extended as far upstream as present-day Moab, Utah. Remnants of the Prospect dam are more clearly visible from the west side of Vulcans Throne (see "Optional Side Trip," below).

Directly below where you're standing, Vulcans Forge is visible as a black monolith in the middle of the river, upstream from Lava Falls rapid. It is believed to be a volcanic neck, a shallow intrusion of basaltic magma.

OPTIONAL SIDE TRIP

For those who would like to see more of the volcanic features in the vicinity of Toroweap, the trip to the west side of Vulcans Throne can be very rewarding. *If you wish to make this side trip, you must have a high-clearance vehicle.* **Although passenger cars can make it to Toroweap Overlook in good weather, the road to the west side of Vulcans Throne is deeply rutted and crosses the Toroweap lava flow, making it unsuitable for passenger cars.**

If you decide to make this trip: return to the junction of the unmarked dirt road just 2.7 miles north of the overlook. Turn left onto this road and drive 2.4 miles to the end of the road. There is a small parking area here; this is the Lava Falls trailhead. Park and walk to the canyon's edge.

The Lava Falls Trail descends across the Toroweap cascades from this point to the Colorado River, dropping nearly

3000 vertical feet in a mile and a half. **More of a route than a trail, this extremely difficult hike should not be attempted during the warm summer months, and is an extremely difficult hike at any time of year. If you are interested in making this hike, inquire at the ranger station for further information. Permits are required for all overnight hikes. The volcanic features on the south wall of the canyon are best viewed from the trailhead.**

Clearly visible on the south side of the canyon directly opposite this point is the mouth of Prospect Canyon. Note the dark, horizontal beds of lava at the top of Prospect Canyon; these are the remains of Prospect dam, the highest of the dams in the Toroweap area. Remains of basaltic lava from other dams are also visible on the far wall. The offset along the Toroweap Fault is clearly visible at this point, as well. Note that the left wall of the canyon is about 700 feet higher than the right wall.

Note that portions of Vulcans Throne have slumped into the canyon at this point; the dark red stratified layers of cinders are clearly visible on the slope to your left. Although the precise age of Vulcans Throne is unknown, it is interpreted by most geologists to be one of the younger volcanic features in the vicinity of Toroweap.

Return to the main road, about 2.8 miles from this point.

END

Glossary

Aa - A Hawaiian term for a type of basaltic lava flow whose upper surface consists of jagged, tumbled pieces of lava. Such a surface forms because the brittle outer surface of the flow breaks in response to continued movement of the flow's liquid core.

Advection - The horizontal or vertical flow of material within the Earth. In the context of this book, magma is the flowing material, and flow is driven by the density contrast between magma and surrounding rocks.

Asthenosphere - A zone at the base of the lithosphere in the Earth's upper mantle that is so hot and physically weak that it behaves more like a liquid than a solid. Magma is commonly generated within this zone and rises into and sometimes through the crust to feed volcanic eruptions. This is the mobile layer upon which the Earth's lithospheric plates move.

Breccia - A rock deposit that consists of sharp-edged and angular fragments of pre-existing rocks.

Caliche - a crust of salts (commonly calcium carbonate-rich) that coats preexisting rock surfaces. In arid climates such as the southwestern United States, caliche is a white salt deposited when a wet surface dries.

Cinder - A vesicular fragment of volcanic rock that forms when a blob of molten basalt, typically full of gas bubbles, solidifies during its trajectory or flight in a lava fountain. Cinder is synonymous with *scoria*.

Cinder cone - A conical or round-topped volcano composed of basaltic cinders. Typically forms during a single short-lived eruption. Internal structure is characterized by many layers of cinders that slope away from the top of the cone. Synonymous with *scoria cone*.

Composite volcano - see *Stratovolcano*.

Conduction - The process of heat transfer through solids, from higher temperature to lower temperature, without the movement of the solid matter itself. Compare with *convection*.

Convection - The circulatory (continuous loop) flow of material in response to density contrasts. In the context of this book, convection refers to the circulation of water within the Earth's crust, wherein downward flowing and relatively cool and dense water becomes heated by adjacent rocks at depth, and thereby becomes less dense and buoyant such that it flows upward. Some of this circulating water may leak to the surface at hot springs.

Dendrochronology - The study of annual growth rings in trees as a means of determining age and environmental conditions.

Endogenous - The type of lava dome that grows by expansion from within and develops a concentric, "onion-skin" arrangement of flow layering in the process.

Exogenous - The type of lava dome that grows mostly by multiple periods of extrusion of lava that add new layers to the outside of the growing body.

Extrusive - Rock formed from lava that cools and solidifies at the Earth's surface. All volcanic rocks are extrusive.

Graben - A down-dropped block of rocks, bounded by steeply dipping faults on both sides.

Harmonic tremor - Continuous ground vibration. Such vibration may continue for several minutes to tens of minutes, and typically occurs at a frequency of about 1 to 2 cycles per second. In volcanic regions, harmonic tremor is most commonly interpreted as an indicator of the flow of magma within the Earth's crust. Also called volcanic tremor.

Intermittent river - A river that flows only certain times of the year, generally when it receives water following heavy precipitation.

Intracanyon flow - The type of lava flow that enters a drainage valley and flows downstream.

Intrusive - The type of rock formed from magma that cools and solidifies beneath the Earth's surface (for example, granite). Also called *plutonic*.

Joint - In volcanic rocks, a surface along which the rock breaks due to shrinkage in volume that accompanies cooling from lava to solid rock.

Lahar - The deposit that forms when loose rock fragments, of many sizes and shapes, slide or are washed down the flanks of a volcano. Lahars are most commonly associated with stratovolcanoes.

Lava - See *magma*.

Lava dome - A steep-sided, dome-shaped volcano that forms when highly viscous magma piles up over its eruptive vent. The core of a dome consists of lava, in contrast to loose volcanic debris, and is commonly enveloped by loose-and-tumbled fragments of lava that form when internal expansion fractures the solid and brittle outer surface of the dome.

Lithospheric plate - A piece of the outermost brittle rind (from 4-25 miles thick) of the Earth. The lithosphere (which includes the crust and the rigid portion of the upper mantle) is broken into about a dozen pieces that move about over the asthenosphere. Also called tectonic plate.

Magma - Naturally occurring molten rock beneath the Earth's surface. Magma commonly contains some solid material and dissolved gases. Called *lava* once it erupts onto the Earth's surface.

Monogenetic - Said of a volcano that forms during a single, short-lived eruption (up to a hundred years or so).

Pahoehoe - Hawaiian term for a type of basaltic lava flow whose upper surface is smooth or ropy, but not fractured and broken.

Perennial river - A river that flows year around under normal precipitation conditions.

Planetary accretion - The process by which planets of our solar system first formed. Bits of interstellar dust and gas fell together through the force of gravitational attraction to form the planets. Heat was produced during accretion as debris collected into discrete planetary bodies.

Plate tectonics - The widely accepted theory that plates of the Earth's lithosphere are in constant motion relative to each other and interact along plate boundaries to cause earthquakes and rock deformation.

Plutonic - See *Intrusive.*

Polygenetic - Said of a volcano that forms during several eruptions that occur during a period of thousands to hundreds of thousands of years.

Scoria - See *Cinder.*

Scoria cone - See *Cinder cone.*

Seismometer - A scientific instrument that measures earth shaking. A network of seismometers is needed in order to locate earthquakes and determine their magnitudes.

Shield volcano - A broad, low volcano built up from many lava flows of basaltic magma.

Silica - Name given to the chemical compound silicon dioxide (SiO_2); quartz. Also used in the reporting of chemical analysis of a rock.

Silicate liquid - A liquid that consists primarily of a network of pyramid-shaped arrangements of one silicon (Si) atom surrounded by four oxygen (O) atoms. The pyramids are commonly linked by atoms of aluminum (Al), calcium (Ca), sodium (Na), potassium (K), iron (Fe) and magnesium (Mg). Most magmas are silicate liquids.

Skylight - An opening in the roof of a lava tube, caused by local roof collapse.

Stratocone - See *Stratovolcano.*

Stratovolcano - An upward-steepening, cone-shaped volcano that is built up of alternating layers of lava and cinders, mostly of andesitic composition. Lahars are commonly present along the mid to lower flanks and around the base. Synonymous with *composite volcano* and *stratocone.*

Tuff - Consolidated volcanic ash.

Unconformity - A break or gap in the geologic record, where one rock layer is overlain by another of substantially younger age. The unconformity represents the missing rock record.

Vesicle - A void or cavity in a volcanic rock that represents a trapped gas bubble. Vesicles are typically $1/2$ inch or less in diameter.

Viscosity - A measure of a fluid's internal resistance to flow. High viscosity equates to high resistance to flow. The viscosity of any given lava is (at least partially) dependent upon its composition.

Volatiles - Those constituents of magma (such as dissolved water and carbon dioxide) that tend to escape as a gas when magma moves upward in the crust, to areas of lower confining pressure.

Index